Renewing the Countryside

WASHINGTON

This Complimentary Copy
Provided By:

Northwest Area
Foundation

and

Sustainable Northwest

RENEWING THE COUNTRYSIDE — WASHINGTON

EDITOR:

John Harrington

WRITER:

Ingrid Dankmeyer

SENIOR PHOTOGRAPHER:

Jim Anderson

ART DIRECTION AND DESIGN:

Brett Olson

A PROJECT OF

Sustainable Northwest
in partnership with Shared Strategy for Puget Sound, Farming and the Environment, and
Washington State University Center for Sustaining Agriculture and Natural Resources

PUBLISHED BY

Renewing the Countryside

RENEWING THE COUNTRYSIDE — WASHINGTON

Editor	John Harrington
Writer	Ingrid Dankmeyer
Senior Photographer Photographers	Jim Anderson John Burks Vaughn Collins James Dooley Ashlyn Forshner Gary Gehling John Harrington Martha Jordan Tom & Pat Leeson Tori Lenze David Perry Richard Sakuma Terri Smith Sunny Walter Sam Walton
Art Director	Brett Olson
Project Coordinator	John Harrington
Series Editor Assistant Editors	Jan Joannides Andi McDaniel Michelle Wilwerding
Sponsors	National Fish and Wildlife Foundation Northwest Area Foundation Walton Family Foundation
Printer	Friesens, *Printed in Canada*
Paper	105 lb. Reincarnation Matte, 100% recycled, 50% post consumer waste, New Leaf Paper, San Francisco
Cover Photography	Cover: Jim Anderson Spine: Brett Olson

ISBN-0-9713391-7-1 (hardcover); ISBN-0-9713391-8-X (paperback)
Library of Congress Control Number: 2005901768

First Printing

TABLE OF CONTENTS

FOREWORD

Most of Washington's citizens live in urban and suburban centers and their sense of connection to rural places has faded. Indeed the health and prosperity of most rural landscapes and communities has been declining for decades. Sadly, this is a national—even a global—trend. Allowing this trend to continue unhindered and unabated is to risk the livability of Washington for future generations.

Rural Washington offers more than places to enjoy for their amenity values. It is the source of the water we drink. It is the generator of the clean air that we breathe. It is the habitat of fish and wildlife that provide recreational opportunity and a cultural identity for our state. It grows much of our food and many products that we use to build and furnish our homes. It is the regulator of our climate. It is also the place that nourishes our souls, and creates lasting memories for all who live there or visit.

The future of Washington lies in a shared sense of stewardship for people, communities, and the lands and waters we have inherited. We must take on this challenge together because nature does not respect political boundaries. Watersheds transcend counties. Rivers and streams flow through strings of cities, farms, and tribal lands. We are just starting to learn more about how species and ecosystems interact and about the short- and long-term consequences, most unintended, of people's actions.

Happily there are many people in rural Washington who are charting a new course for all of us by seeing themselves as a part of a larger ecosystem. They are the innovators of that stewardship imperative. By sharing their experience, they can also be our teachers. *Renewing the Countryside: Washington* tells the stories of just a few of them. Each of them inspires a can-do attitude that should excite and motivate us all.

We live in a region that has experienced its share of political and environmental conflict. These "renewers" have moved beyond conflict to coexistence and cooperation. They understand that the future requires that they work with—not against—nature, and that profit must be measured in more than money. Relationships, trust, meaningful work, and community are all part of that expanded definition of profit. So is protecting and restoring nature. Even so, most of the examples in this book demonstrate that it is possible to be profitable through good land stewardship.

These "renewers" are not just renewing the countryside. They are renewing urban–rural linkages. They are renewing our hope that each of us—as producers and consumers, citizens, public servants, teachers, students, manufacturers, and community members—has a role to play.

We will all learn from the messages embedded in each of the stories contained in this wonderful volume. Our state is blessed to have so many enlightened and dedicated citizens. We thank them for their work, their commitment, and their willingness to share their stories with all Washingtonians.

Bill and Jill Ruckelshaus, Medina, Washington

INTRODUCTION

The words and images in *Renewing the Countryside: Washington* are about people achieving extraordinary results that give the rest of us hope. They are about peoples' intimate relationship to their surroundings and the bounty that their commitment produces. They are also about communities that define themselves through the values and spaces they share in common.

In these pages, you will see why Washington State can be known for its commitment to environmental stewardship and community vitality. For those featured, these are not abstract concepts. The forty-three examples profiled in this book highlight the relationship people can nurture with the land, air, water, wildlife, and others they live and do business with. This relationship forms an intricate community—a very real and personal web.

Through these stories, we can appreciate how much our existence depends on well-managed natural resources, healthy watersheds, and vibrant communities. We can also realize how much we benefit from the labors of stewards like these. By working in concert with natural systems, these people and their enterprises produce life-giving foods, conserve resources for future generations, enhance the economic viability of rural communities, and protect the needs of people, fish, and wildlife.

In a story, any ending is possible. These stewards show us how to refuse an ending that does not include all three ingredients of sustainability: economic prosperity, vibrant social networks, and a healthy environment. Through these real-life examples, we see our happiest conclusion—a place where people are taken care of as they take care of the land.

As publishing partners, our four organizations have brought different objectives and points of view to this project. The value of working together on this book has been the opportunity to share and learn from each other's expertise and familiarity with the broad range of issues affecting Washington's rural communities and landscapes. We hope this book inspires you, as it has us, to make the story of Washington even richer and worthy of re-telling. We live in a mythic place, and we are at our best when we share the responsibility for keeping it that way.

Martin Goebel, President, Sustainable Northwest
Chris Feise, Director, Washington State University Center
for Sustaining Agriculture & Natural Resources
Jim Kramer, Executive Director,
Shared Strategy for Puget Sound
Joan Thomas, Board Vice Chair,
Farming and the Environment

COMMUNITY

In Washington, as in much of the West, rural communities are often established around natural resource-based industries such as agriculture, ranching, fishing, and timber. As settlement and development contribute to a decline in the health and abundance of those resources, the vitality of communities dependent upon that natural wealth also suffers.

Wonderful things can happen when people reach out across property lines and cultural divides to improve their community: restoring watersheds, establishing new traditions, attracting visitors, engaging diverse participants, and reviving the local economy.

In this chapter we are reminded of the enduring importance of communities of place, where residents share an interest in the vitality of a particular landscape and the people who call it home. These stories demonstrate how collaborative community efforts can lead to creative and enduring solutions to local challenges.

DUNGENESS RIVER WATERSHED RESTORATION

"Every river has its people" has long been a saying among the native people in the northern reach of the Olympic Peninsula. In the past two decades the Dungeness River's people in northern Puget Sound have avoided litigation by collaborating to resolve water use disputes, working together to restore the river that sustains them.

The Dungeness River is one of the steepest rivers in North America. It originates up around 7000 feet in the regal Olympic Mountains and descends a dramatic 4000 feet over its first four miles. Its lower ten miles flow through the Sequim-Dungeness valley, a uniquely arid part of the mostly sodden Olympic Peninsula. A rain shadow cast by the mountains means that a scant 16 inches of rain falls annually where the river lets out into the Strait of San Juan de Fuca, close to the Canadian border.

Archeological evidence shows that people, attracted to an abundance of salmon and other coastal resources, have inhabited this 27-square mile watershed for as many as 11,000 years. But as human population has increased in the last 150 years, the landscape has changed dramatically. An 1855 treaty compelled the indigenous S'Klallam Tribe to give up their claim to 438,000 acres of land, but they retained the treaty rights to fish, hunt, and gather on the north Olympic Peninsula. At that time European settlement began in earnest, bringing logging operations that exported timber down the Pacific coast and farmers who diked the lower river out of its floodplain and began building what is now over 170 miles of irrigation ditches in the lower Dungeness Valley.

In 1874 when the S'Klallam Tribe was threatened with relocation to a reservation, a handful of families pooled 500 dollars in gold coin to purchase 210 acres of land along Dungeness Bay; they called their community Jamestown. Today their descendants comprise the 525 members of the Jamestown S'Klallam Tribe. The Tribe's natural resources department is charged with protecting the fish and wildlife to which the 1855 treaty gave them access. Reversing dwindling salmon runs are a major priority. In 1963, the first year they were counted, there were 400,000 pink salmon in the Dungeness River; by 1981 the population was down to 2900. Currently eight stocks of Pacific salmon return from the ocean to spawn in the Dungeness River. Two of those—chinook and summer chum—are listed as "threatened" for this region under the Endangered Species Act.

Ann Seiter started working as the Tribe's natural resources manager in the mid-1980s. She remembers a Tribal council elder saying to her, "Please do something about the Dungeness, the fish runs are going down and you can tell that there's no water for the fish." Ann started by researching water rights and found that the local farming community had legal access to far more water than the Dungeness could provide.

"There were nine different irrigation districts and companies in the valley at that time and they didn't talk to each other," says Ann. "There was no central association, so if we wanted to ask for voluntary conservation, we had to call all nine entities and we'd get varying levels of cooperation." In those early years it was very

COMMUNITY

contentious with the farming community. When farmers were asked to conserve water for salmon, they felt very threatened.

Ann remembers how local farmer, Roger Schmidt, tried to get fellow farmers to talk to the Tribe early on and the criticism he received for his efforts. Nonetheless, he succeeded in bringing the irrigators in the valley together under one umbrella, the Dungeness Water Users Association. "Roger got folks organized, and that was a big step," recalls Ann.

In 1987, a severe drought reduced the Dungeness River to a trickle you could step over, and pink salmon had to be captured and trucked upstream. At an antagonistic meeting between the Tribe and the

water users, it became clear that neither side had data about how much water was being withdrawn, nor how much salmon needed. "You can shout about how many fish there are and how much water is in the river, but unless you actually have data you're just shouting about who has a right when you don't even know how much of the resource you are using or that you need," notes Ann.

Subsequent measurements revealed that a whopping 82 percent of the Dungeness River was being diverted for irrigation during the dry season, which didn't leave much for salmon. "For the farmers and the Tribe alike it was a startling statistic," Ann says. "And while it is hard to explain to some of the old timers why we can't get by with the old practices, I think the leadership of the irrigation districts understands."

The push to work together to improve water levels in the river came from other quarters as well. The Department of Ecology made the case to the farmers, advising them to avoid litigation if they could. For its part, the Tribe preferred to avoid attorney fees, and they knew litigation could take years while the salmon runs continued to decline.

When another Sequim farmer, Dave Cameron, was elected to the county board of commissioners, he quickly made good on a campaign promise and formed the Dungeness River Management Team, bringing the Tribe, irrigators, property owners, conservationists, and public agencies together. According to Ann, "That was extremely constructive. We started meeting monthly right off the bat, developing personal relationships to talk about what the problems were, to begin to define the technical questions that needed to be answered."

The Dungeness River Management Team worked together to forge a groundbreaking voluntary agreement. The irrigators agreed to withdraw no more than 50 percent of the river's flow in perpetuity, even though they are legally entitled to much more than that. The Tribe agreed to help with conservation efforts, and to work to address other impacts on fish habitat. Water conservation was not the only problem—dikes, land use, and forestry also has detrimental effects—and the farmers wanted that acknowledged.

The century-old irrigation system in the valley was in need of some serious repair, and the Tribe helped the farmers secure over one million dollars in grant funds for improvements. Enclosing ditches and lining pipes has allowed more water to stay in the river, while also giving the farmers a more reliable, easier to maintain irrigation system. In 2001, another drought year reduced the river to levels that were similar to those in 1987, but this time, irrigators withdrew just 33 percent of the Dungeness's flow.

"The big conflict between fish and irrigation really occurs around the first of September," explains Ann, "because that is when it's the end of the irrigation season—it's a last watering for the farmers and it's the peak of chinook spawning. We've been trying to focus a lot of effort on water conservation at that time."

Ann reflects, "My first ten years on this issue we spent a lot of time focusing on water conservation because it was a no-brainer; it was obvious we had a problem when you could step across the river in the summer. The last ten years there has been a lot of effort on floodplain issues, a lot of studies being done, and a lot of property is being purchased along the riparian corridor through mostly public funds and the local land trust."

"Property owners in flood prone areas are being bought out," Ann explains. "Most of the property owners want to get out, so it is mutually beneficial." As property in the river's historic floodplain becomes available, it will be purchased; buildings and septic systems will be removed along with some dikes, so that the river's original meanders can be restored. "We are making a big effort to minimize new construction close to the river, particularly where people are threatened by flooding and where fish habitat is also a concern," says Ann.

Rapid growth and development of this sunny corner of the Olympic Peninsula is perhaps the most pressing current concern for the Tribe and farmers alike. Over the last 30 years, the number of retirees living in the area has tripled. "New development is still a threat," adds Ann. "We have concerns about whether county regulations are stringent enough. I am particularly concerned about urbanization encroaching on the river." With this new influx of residents to the valley, there will continue to be more people to engage in sustaining the vitality of the Dungeness River.

LEFT FOOT ORGANICS

The mission of Left Foot Organics extends well beyond that of most market farms that provide fresh produce to subscribers and other customers. Ann Vandeman, the director of this three-year-old non-profit venture, thinks as much about the experience she provides her farm hands as she does about what comes off of the fields.

Left Foot Organics offers adults with developmental disabilities the opportunity to grow organic produce while strengthening their social skills and self-reliance. Ann explains, "Our purpose is to provide

employment and life skills training to people with disabilities and to use these activities as a means of educating the public about people with disabilities and the need for inclusion." Left Foot Organics demonstrates the contribution that people with disabilities can make to their community through concrete products people want and that have value: certified organic vegetables and berries.

For Ann, this venture is very personal. She explains, "I started this because I have a daughter with Down Syndrome. I was working for

the Department of Agriculture in Washington DC; when she was born, I wanted to come home to the Northwest. I wanted to get out of the office and into the fields actually producing something other than reports about data; I wanted to be working with people with disabilities." The field of horticultural therapy, using horticultural activities to achieve therapeutic results for people with special needs, combines Ann's two passions. After volunteering for a similar project in Washington DC, Ann returned home and started hoeing in 2001.

Left Foot Organics leases two acres south of Olympia, within sight of Interstate 5. Already, the farm employs ten people with developmental disabilities, three additional part-time staff, Ann, and three volunteers. Recently, they have added a youth employment program in collaboration with a youth service agency in town. Ann says, "That is really important for us because it integrates our workforce. We are mixing up these typically developing youth and youth with disabilities, exposing both of those groups to each other and modeling the kind of community we want to create."

Ann describes her vision: "We want the community that these folks live in to be welcoming them—in employment, in school, in social life, in public life—and so we're creating that kind of environment here on the farm where everybody is working together toward common goals and contributing to the best of their ability." Ann explains that it doesn't matter if one person can bunch chard and another person can dig potatoes. Everybody does whatever they can do and they do their best. The result is a product that people in the community want to buy and so they are compelled to interact

with someone with disabilities that they might otherwise ignore. Ann says, "Through this transaction—someone buying snow peas, or whatever, from a person with disabilities—we're creating a relationship that acknowledges the contributions of people with disabilities in the community."

Left Foot Organics concentrates on direct marketing because it creates those important social interactions. Says Ann, "We sell shares, and our customers become invested, and they come out for potlucks!" The farm feeds 40 subscribers off of just two acres, providing weekly boxes of vegetables, herbs, and flowers. Left Foot produce is also sold at several area farmers' markets, as well as in a handful of local stores. Last year a local school district bought potatoes and carrots, and Ann hopes to see that partnership grow. "Selling our produce to the schools doesn't get us as close to the consumer as I would like," she remarks, "but the schools have surplus land that they have offered to local farmers." At some point Ann would like to see students with developmental disabilities working together with their typically developing peers on school land to produce food for the students.

Over 50 varieties of vegetables—from arugula to zucchini—grow in soil that is visibly rocky. Left Foot Organics tills part of an old dairy farm, and the office is in the old milking parlor. Says Ann, "None of this land had been in production for a long time, so just coming in from scratch and trying to get the soil into shape has been a huge challenge." Fortunately, there hadn't been chemicals applied for some time so they were able to get organic certification immediately. Manure and cover crops are helping to build the soil fertility, but the weeds and rocks remain formidable. "I feel bad about running back and forth over the field with heavy equipment because it compacts the soil," Ann laments, "but weed control is one of our biggest problems, so we have to. And the rocks just destroy our equipment; that's been really difficult."

Farm manager Beth Leimbach acknowledges, "This land has been a tough teacher. The soil fertility, rocks, and weeds are really a challenge on this particular farm, and sometimes added to that is this more complex staff and having enough people who have enough time to get the job done." But Beth has learned as much from the people with whom she works as from the land. "My learning," Beth says, "is that change may not happen for people with developmental disabilities as fast as I'm used to seeing change. It takes a season and it's so organic that way, just like the farm. You can't learn farming in one year—you have to do it several years in a row."

Ann hopes to grow Left Foot Organics to provide more opportunities for people with disabilities. "I'd like to have ten acres and pasture and some small animals," she says. "If we could do sheep and spinning and weaving, then we'd have people working year round. And if we had some chickens and eggs, we could provide a more varied experience for the growers." A greenhouse built last year, with time and materials donated from local carpenter and electrical unions, means they can start their growing season a month earlier this year.

Ann would like to see a resurgence of vocational farming programs. "All the large mental institutions, prisons, and residential facilities for people with developmental disabilities used to have working farms," she explains. "Then there were issues around compelling people to work, and so they were all shut down. They kind of threw the baby out with the bathwater because these activities are valuable to people, and if they could find a way to do it so it isn't coercive or somehow exploitive, why not bring those back so people have something productive and meaningful to do?"

A volunteer crew from a local minimum security prison has put in time at Left Foot Organics. Ann says, "They are good guys, hard working, and they like helping and knowing they are doing something really valuable—like putting up bean trellises—things that take a lot of people over a short period of time to accomplish. It's been really nice to have their help and I think more help of that kind would be valuable."

On this farm, the fresh produce is a valuable byproduct of the real work. Says Ann, "Farm activity is good for people with disabilities. It has a lot of different benefits: fine motor development, improved coordination, improved cognitive function, and learning social skills by working as a team and interacting with the public." She continues, "Like all of us, when people with developmental disabilities have meaningful work that gives them a sense of accomplishment and purpose, and brings opportunities to contribute to their community and be recognized for that, it improves their quality of life."

MADE IN THE METHOW CO-OP STORE & COMMUNITY KITCHEN

Downtown Twisp is less of a tourist destination than the Old West-styled town of Winthrop just up the road, but those who take the time to mosey down Twisp's quiet main street may discover more about the Methow Valley than those who simply speed past. The Made in the Methow Co-op store features an appealing, eclectic mix of local foods and crafts. Refrigerators stocked with tamales, specialty cakes, grass-fed beef, and soy sausage are nestled among the wares of local herbalists, musicians, potters, and jewelers.

This venture began with the food. Local food production in the Methow Valley has been well supported for years by a large, popular weekly farmers' market. In 1997, some of those vendors began talking about increasing food production capacity in the community. There was a lot of produce being locally grown, but not much of it was being sold as higher-value products like jams and salsas. By law, food products offered for sale must be produced in a licensed commercial kitchen, but such a kitchen is costly to build. So a handful of vendors and folks interested in community development began discussing the idea of creating a shared kitchen facility in the Methow Valley.

After forming a sponsoring community organization—the Partnership for a Sustainable Methow—and surveying potential kitchen users, state funds were secured to develop a business plan and build a kitchen in downtown Twisp. Joyce Campbell is a farmer, soy foods producer, and one of the early leaders of the kitchen and the Partnership. She remembers, "We received a two year grant that gave us a part-time manager and all of the equipment. The grant was matched by the community through donations and all kinds of help." The kitchen opened in 2000, and salsas, cakes, and jams were added to the mix of locally made and sold food products. Adding a walk-in cooler provided much-needed cold storage space for farmers.

In the kitchen, Ann Wagstaff makes hot sauces and salsas that she sells in local stores and at the farmers' market. She remarks, "It just helps a lot not to have to do that huge outlay of cash to get your own commercial kitchen. There are a lot of requirements to do that, and if there is community effort and community donations to get it going, it eliminates a huge stumbling block for a lot of small businesses." Ann's business is a part-time endeavor so the kitchen has been perfect for her. She explains, "I can rent the kitchen by the hour and it has all the equipment I need. It is also excellent for people who are experimenting with a product."

Today the kitchen and store are incorporated as a cooperative business, and members include farmers who rent storage space in the cooler, folks who use the kitchen to bake pies and pizzas to sell at the weekly farmers' market, and local artists and food producers who sell their wares at the store. Kitchen users pay an hourly rate for use of the facility, and those who sell their products in the store volunteer their time and/or pay a commission on sales. Local subscription farms rent space in the cooler to keep produce fresh for customer pick up, along with pear and apple growers who need long-term storage. The facility is also used by caterers, home canners, and local event organizers.

Two businesses have successfully graduated from the shared kitchen to their own facilities—Joyce Campbell's Methow Valley Foods and Salyna's Specialty Cakes. Cooperative member Sara Hartzell explains, "Currently we have about 22 merchants who sell products in the store. The store has carried the kitchen since those two businesses incubated and moved out. We are hoping for more businesses to move in here and get started."

"It's really a labor of love," Sara continues. "The vision everybody has for the Methow Valley is that it can support residents and help people start businesses. The big income for the region is tourism, so a lot of people in our area don't have jobs in the wintertime. They go on unemployment year after year, and that is a strain to the economy here. This is one way everybody can help out."

While most of the store's customers are tourists, locals come in regularly for the frozen soups and beef. A web page is in the works so people who have visited the area can access a piece of the Methow Valley from anywhere. But web visits are unlikely to capture the experience of browsing among the wreaths and knitted hats hanging from branches, the stained glass pieces glinting in the window, the smell of herbs and scented lotions, and the blackboard on the back wall describing the bounty of the Methow.

Over time the kitchen has expanded its scope to serve a broader base of community needs. From the beginning, the founders had envisioned that the Partnership would encompass more than agriculture, supporting local businesses across the board. Its mission is "to encourage and support sustainable economic development in the Methow Valley through activities that preserve the rural environment and quality of life." Partnership executive director, Sue Koptonak, remembers when the kitchen opened: "We were experiencing a lot of job loss. The timber and agricultural markets were changing. Downtown Twisp had a lot of empty storefronts."

The kitchen was designed as a business incubator that would support entrepreneurs in the Valley, food-based and otherwise. In 2001, the front of the kitchen, on Twisp's main street, was transformed into the Made in the Methow retail store. There had always been a vision of being able to sell food products out front that were made in back, but Ann explains there wasn't enough variety of food items made in the Valley to warrant the store. She says, "So we opened it up to anything made in the Methow, and now it's art and stained glass and hand crafted items—and food!"

OTHELLO SANDHILL CRANE FESTIVAL

Not so long ago, Othello was just another sleepy agricultural town in an untrammeled part of the state. The lake-studded Columbia National Wildlife Refuge nearby was known more for hunting wildlife than viewing it. Today, the chance to catch sight of the red-capped, leggy sandhill cranes that stopover here on their seasonal migration between Alaska and California has attracted a whole new flock of visitors to the area.

The wildlife refuge was established as part of the Columbia Basin Irrigation Project that diverts water from Grand Coulee Dam on the

Columbia River. Refuge biologist Randy Hill says the initial land was secured because no one else wanted it. It was wet, and unproductive for crops. Over time, additional areas for migratory bird habitat were acquired, and the refuge grew to include 23,000 acres. The numerous lakes and wetlands on the refuge are kept moist by tiny seeps through the dam on the Potholes Reservoir, just south of Moses Lake, and the irrigation canal that forms much of the east boundary of the refuge.

The refuge includes farmland that is cultivated to provide a year-round food supply that helps migratory birds maintain healthy populations. Randy explains, "During the fall migration when the birds are heading south, we provide crops they can eat so they can get to their destination in reasonably decent shape." If the birds are at the refuge in the winter, they can maintain their body weight and not freeze or starve to death. Randy adds, "In the spring, when they are heading to their breeding grounds in southeast Alaska, we want them to take up as much fat reserves as possible, because once they lay eggs, there is a lot that goes into sitting on those eggs for days on end."

Ten years ago, the refuge's service to migrating birds was not foremost in the public's mind. Randy says, "We have always had a lot of users of the refuge, but people have thought of it as a place to go fishing or hunting. A lot of locals knew it as a place to get off the beaten track and unwind, but it never really was seen as a place to go see migratory birds." Randy himself didn't initially realize the potential of the avian attraction on the refuge. He knew there was a sandhill crane migration going through the area, and had taken a few people out to see the cranes, but didn't think a lot about it. Randy reflects, "But word kind of gets out because they are a big showy species. They are communal, gregarious, and hang out in pretty big flocks sometimes."

A birder in Moscow, Idaho, heard about the cranes and drove over to the Columbia National Wildlife Refuge to see them for himself. When he wasn't able to find any, he tried another tactic: he offered to lead a crane-viewing field trip through the University of Idaho's adult education program. With Randy as their guide, a van full of Idahoans were able to see cranes roosting and feeding in and around the refuge in the spring of 1997.

At this time, Greater Othello's Chamber of Commerce was looking for a way to generate more traffic for local businesses. Randy explains, "The Chamber of Commerce realized that other than agriculture, about the only thing the Othello area had to offer was outdoor recreation and a lot of that was concentrated on the refuge."

So the Chamber and the refuge manager, the US Fish and Wildlife Service, joined forces to create the first Othello Sandhill Crane Festival in the spring of 1998. The International Crane Foundation offered support, and the Washington State Audubon Society also got involved. Randy remarks, "That was excellent promotion because there are 25 Audubon chapters throughout the state and they all have newsletters. Audubon members are the targeted audience for a wildlife festival like this." Local papers and birder's websites provided additional free coverage, and that was about the extent of the advertising.

More than 400 people participated in the first festival, and Randy estimates as many as half of them were from the Othello area. The school district was one of the event partners so the high school facilities were used for hosting activities and school buses were used for the bird tours. That first year some of the challenges of balancing the interests of wildlife and tourists became apparent.

Randy recalls, "We realized that it was difficult to take people out every hour to view cranes because the birds are bimodal: they go out to the fields to feed at dawn and then go back to the roost area and sit there for the rest of the afternoon. They leave again in the late afternoon or early evening to feed in the fields again before going back to their nighttime roost." When they couldn't see the birds in the fields, they took the buses to the roost to see them. Randy says, "I realized that first year that we should not have gone into the roost. People got great looks—the birds got up, they were flying and they made a lot of noise, but the reason the birds got up is that they were disturbed."

The first year set the stage and was an opportunity to build awareness in the local community. Says Randy, "People sort of knew there were cranes around but didn't really know anything about them. We opened a lot of eyes locally." The festival organizers also realized they needed more local people involved in running the festival. Randy explains, "The first year we had wildlife guides on the buses, and they were getting a lot of questions about local agriculture as they drove through the fields. The next year we had wildlife and agriculture guides on each bus—locals that could talk about irrigation systems and different crops."

Othello's Sandhill Crane Festival has grown into a three-day event that attracts 1500 people. In addition to the crane viewing, visitors can participate in a wide array of specialty tours focused on other bird species and the unique geology of the area. Free lectures feature topics like owls, falconry, shrub-steppe flora and fauna, and the Missoula floods. Young visitors can make crane masks or fold origami renditions of the birds. An author's forum includes local naturalist writers and appearances by luminaries such as Peter Matthieson, who wrote a book recently on the 15 crane species of the world.

"When we started the refuge, fishing was king," says Randy. "I don't think fishing is king anymore. Not only has fishing declined somewhat, but with the publicity from the festival, people come to see the cranes, they go through the refuge, they

find other neat stuff, they tell other people." Changes in crop management on the refuge has also brought more cranes. "We get a crane flock that builds to over 5000 at one time. That is over 20 percent of the population in the entire flyway," notes Randy.

Interest in seeing cranes has spilled over to other weekends, and tourist visits have increased throughout the migrating season. Hotels are happy with the business they get on other weekends and during the week as well. Explains Randy, "We're at a point now where we can't do it all in one weekend anymore. The festival will be on one weekend, but we are going to have to offer crane viewing on other weekends as well because the crane viewing resource is limited." Randy concludes, "What started as a crane festival has expanded into broader ecotourism. It has worked out really well for the community!"

SEQUIM LAVENDER FESTIVAL

In Sequim, a shared dream of sweet-smelling purple fields of lavender has changed a town's character and landscape. In nine short years, Sequim has gone from relative obscurity to national renown as The Lavender Capital of North America.

"Lavender now grows on land once occupied by thousands of dairy cows," says local Lavender Festival historian, journalist, and crafter, Betty Oppenheimer. "During the first half of the 20th century, hundreds of dairy farms thrived in the valley, but today, only a handful of full-scale farms remain. Sequim lost 75 percent of its agricultural base in five decades. Retired farmers have seen the value of their land rise exponentially if they sell it for residential development rather than for farming." This scenario is familiar to many communities across the west, as is the result: farmland is being subdivided and developed.

But Sequim is also unique. The Sequim Dungeness Valley lies in a "rain shadow" created by the high peaks of the Olympic Mountains, thus the clouds part more frequently here in the winter than elsewhere in western Washington, attracting folks looking for a bit more space and a bit more sun. "Over the past 30 years, Sequim has become a haven for retirees," says Betty. "Golf courses and new homes have replaced much of the farmland, and many who own acreage no longer farm it."

Betty continues, "As the town grew, its beauty was often obscured behind shopping centers and recreational vehicle parks. The idea of growing flowers began as a vision of rediscovering the beauty of the remaining land, in the hopes of creating a community identity which would ultimately preserve it." Residents of Sequim, many of them newcomers, joined together to reinvent and revitalize the area's agricultural heritage.

As farmers subdivided and sold their land, gardeners and people looking for a quieter lifestyle moved to the area. Betty laughs, "Mowing five acres of grass and weeds soon lost its romantic flavor, and many of these new landowners were seeking a better way to utilize the land." The idea of growing lavender in the Sequim Dungeness Valley was first discussed in 1995 by an informal group of farmers, gardeners, and community leaders. Betty explains, "Their objective was twofold: to preserve Sequim's agricultural land and to create a tourist destination where people would come to learn about farming, and contribute to the health of the area. The group chose lavender because it thrives in sun and dry soil, and Sequim, because of its unique prairie microclimate, offers both."

The earliest lavender fields were planted in1995. By 1997 when the first Lavender Festival was held, there were seven lavender farms; today there are close to 40. A Saturday open air market was launched in 1996 to bring together local craft, produce, and, of course, lavender sellers. "The first Celebrate Lavender Festival featured farm tours, a downtown market, and an evening dance," recalls Betty. "The festival has grown into a long weekend of tours, a downtown craft fair with over 100 booths, activities, and entertainment." In 2003 the festival attracted 28,000 visitors and brought an estimated $2.5 million into town in one weekend.

Because different varieties of the plant bloom throughout July and August, the lavender season in Sequim lasts all summer, and many farms stay open well before and after the festival.

"People now associate lavender with the small town of Sequim, just as everyone had hoped," says Betty. The economic impact of Sequim's new focus has been far reaching. "Each farm looked for a marketing niche," Betty explains, "from you-pick flowers on the farm, to supplying the wholesale flower market, to making products. And since many farmers were simply too busy to make their own products, and many crafters had no interest in farming, cottage businesses formed to fill a need. There are currently several dozen businesses, separate from the lavender farms, who make at least part of their living making soap, baked goods, candles, sewn products, plant starts, aromatherapy, and cosmetics. Other kinds of support businesses grew with the lavender industry too. Marketing firms, printers, and web site designers all found new customers in lavender farms."

The community has been revitalized by its new identity. "We've learned to be open to new ideas, to welcome strangers, and to believe in our ability to create new things," says Betty. "We've learned that it takes all kinds of people with all kinds of skills to change a community identity."

Of course there have been some challenges along the way. "The tensions between farmers who aggressively pursued markets and farmers who wanted a more cooperative spirit were very real," explains Betty. "But as people explored market options and personal preferences, each farm found its niche." Other issues—like product standards, pricing, whether or not to grow organically—have arisen as the lavender industry has grown. Betty notes that each year some folks come to Sequim thinking they can invest in land, plant lavender, and it will sell itself. "Not so," she says.

And while the Sequim Dungeness Valley is now dotted with aromatic fields of purple, farmland is still threatened. "Agritourism is a double-edged sword," explains Betty. "Lavender brings in lodging and restaurant customers, but it also crowds the town, and gives Sequim high visibility as a great place to live, resulting in more people deciding to move here, and possibly build on what was once farmland. Saving farmland in the face of such a beautiful retirement area is a slow process—one in which each successive wave of newcomers has to be educated anew."

"As a group, it is not the lavender farmers who are saving farmland," Betty acknowledges. That work remains the purview of land trusts and other local activists. "But we have learned that a town can band together to reinvent itself," she says, "even after changes in economy and lifestyle seem to have dealt us a tough blow with the loss of fishing, farming, and logging."

Betty concludes, "The hard work of the lavender pioneers has contributed to the reaffirmation of Sequim's identity as a farming community. Now the community understands the real possibility of economic viability, without depletion of the natural beauty that drew us here. It took visionary leaders, risk-taking entrepreneurs, and dedicated believers willing to volunteer their time to make the dreams real, but we did it!"

TENMILE CREEK WATERSHED RESTORATION PROJECT

Dorie Belisle and her husband John moved from Florida to a 40-acre farm in Whatcom County almost a decade ago to grow Jonagold apples. Dorie acknowledges, "We actually were looking for a quieter lifestyle—just growing apples." It hasn't quite worked out that way. As manager of Tenmile Creek Watershed Restoration, Dorie now spends much of her time talking with neighbors and planning projects to benefit the creek that runs through their farms. While it is not a quiet job, Dorie's unflappable demeanor and positive attitude seem well suited for it.

Dorie and John are passionate about farming and farmland preservation. They grow their apples using integrated pest management and are certified as sustainable by Food Alliance. Dorie says, "After we had been farming for about three years, we got very frustrated about all of the regulations affecting agriculture—a lot of things that just didn't make sense." John serves on the board of the Whatcom Agriculture Preservation Committee, which works to keep farming viable. Dorie remembers, "A few of us from that committee got together and said, 'If we wanted to bring solutions instead of regulation to environmental issues, how would we do it? How can we protect the salmon and also protect agriculture?'"

They came up with a pilot program on the Tenmile Creek watershed that focused on talking with and educating neighbors. Says Dorie, "The main thing is listening to your neighbor and understanding how they use their land, how they farm it, work on it, play in it, why they purchased the property, and what role the creek plays in their long- and short-term goals for their land." Dorie contends that once you understand where a landowner is coming from, you are in a better position to supply that landowner with information about what the salmon and creek need, and you can start finding solutions that work for everyone. With support from the county and the state's Department of Ecology, Dorie became the project manager and the pilot restoration project was off and running.

The Tenmile Creek watershed drains approximately 35 square miles between Bellingham and Canada. It is home to spawning runs of coho, chinook and chum salmon, as well as steelhead and coastal cutthroat trout. "Historically there used to be so many salmon in the creek that the farmers would pitch them out and use them for fertilizer," exclaims Dorie. In recent years, the fish counts have dwindled severely. Dorie says, "I think people are realizing that if we lose farming, and in its place have housing developments, we are going to lose fish. We have the land base for the fish, and with a little tweaking we can make it a very healthy environment for them."

In 2002, Dorie sent a survey to 481 landowners who live on or close to one of the streams in the Tenmile watershed. With the information she received in the surveys, Dorie had the input she needed to get started. For example, 83 percent of respondents agreed that improving stream water quality was important. The survey allowed space for comments that ranged from enthusiastic: "Landowners, let's all get together and do it!"—to irate: "Government should mind their own business"—to flippant: "Kill more seals." Concern was expressed about pollution in the creeks and some people noted that kids were getting rashes from playing in the water. Respondents also stated that a new golf course had adversely impacted water quality and flow. People openly expressed their views, asked questions, and requested more information.

Dorie compiled the survey results and sent them out to all of the landowners. She says, "I went back to them and said, 'Okay this is what you wanted.'" Landowners said they did not want to go to meetings and conferences, but did want someone to walk their land with them, so Dorie arranged outings with small groups of neighbors to walk each other's land and talk about issues and goals. Survey respondents also said they wanted to know more about the history of the Tenmile, so Dorie organized a berries and ice cream social down by the stream where old timers shared their stories of the creek.

Within two years, Dorie made 250 visits to get to know people. She explains, "I might start with a cup of coffee at their kitchen table, just listening to what a family's goals are for the land. The next visit I might bring them information that they ask for on the fish, on the creek, on how the county can help them put in a bridge or remove a fish barrier. The next visit we might start putting a project together." Dorie continues, "The whole thing has

to be based on relationships. My goal is to build a community around the Tenmile Creek watershed so that we start identifying with the area that we live in, because if you don't identify with it, you can't protect it." She also believes that if you remove the regulatory focus and give people incentives, applause, and financial help, they will do what is right.

Dorie has coordinated projects across ownerships on four different stretches of creek, involving 37 families. Most of the effort has focused on voluntary planting of native trees and shrubs along the creeks and in buffer zones. These plantings are site-specific and designed to meet creek and farmer needs. Dorie notes, "We have to get away from thinking in terms of buffer feet and think in terms of what is going to meet the needs of that creek as it meanders through each piece of property."

Dorie offers a cautionary tale about a farmer who had seen three attempts to plant along his stream fail. She explains, "He said to me, 'They planted doug fir in this peat soil and I could have told them they wouldn't grow there, so what are you going to plant?' I said, 'Okay, what will grow here?' The farmer said, 'In the peat soils you are going to need the native shrubs; this site is facing the northeasters that come down out of the Fraser Valley and any big tree is going to blow over by the time it is 15 years old.'" Dorie remarks, "We've got to do what makes sense, and that means talking to the landowners because they know the types of soil on their land and how the wind blows."

Besides listening to landowners, Dorie works hard to help people understand the dynamics of the watershed. She says, "I realize that people don't always understand the connectedness. They don't always understand that what they do in their backyard impacts Portage Bay, and that there are tribes down there that are trying to harvest shellfish, and that they, too, are farmers and what we do impacts their farming down below." She continues, "Our goal is to connect all these farmers, whether you are farming shellfish or you are farming with a dairy up here, we are all connected."

In the time Dorie has been working on this project, four miles of creekside have been planted with native trees and shrubs, over 42,000 seedlings have been planted by farmers in the county, upstream farmers and downstream farmers have broken bread together, and a greater sense of community has been built in the watershed. While the fish runs have yet to bounce back, there are some encouraging signs. "For the first time we have opened up both shellfish beds in Drayton Harbor and Portage Bay after they had been closed for 15 years," reports Dorie. "We are not out of the woods yet, but the dairy industry has stepped up to the plate over the last ten years."

Dorie hopes that this approach to watershed restoration will spread. She believes having a local coordinator who is a good listener makes a big difference. She says, "It has definitely helped that I avidly support agriculture, that I am a farmer, and that I am these people's neighbor." It also helps that Dorie clearly enjoys getting to know her neighbors and she is not easily ruffled.

Dorie tells a final story: "I was out planting trees in the rain with a farmer and he said something about those darn environmentalists. I was wet and cold and I said, 'Look at you, you are planting trees along a stream in the rain with 15 high school kids. If that doesn't make you an environmentalist, I don't know what does.'" She exclaims, "We need to take that term back!"

CONSERVATION

Landowners across Washington State are developing and discovering innovative ways to ensure the resources in their care are managed to meet current human needs without jeopardizing the needs of other species, or future generations.

The resource stewards profiled in this chapter have a long-term perspective. Many of them are working the same land as their ancestors, but with new information and technology in hand. They are revising a few long-standing traditions to work more effectively with natural cycles. Some are independently pioneering new approaches to resource management, while others are taking advantage of technical and financial assistance through local nonprofit and public agencies.

Although the few thousand acres described here represent a fraction of the total land base in Washington, the stewardship practices of these individuals demonstrate what is possible on working landscapes across the region.

JEA Farms, Ltd.

For 30 years John Aeschliman has been pioneering the art of soil and water conservation on 4000 acres in eastern Washington's Palouse region. Take a soil sample just about anywhere, anytime, from the land he farms and a few things are clear even to the untrained eye. Underneath a thick layer of decomposing straw and residue, the soil is soft and dark like well-aged compost. Worms and worm castings can be found in literally every handful of soil. A soil probe driven into dry ground slides easily down to six feet, and at that depth the soil is moist, which isn't what you'd expect in dry country that averages a mere 18 to 20 inches of rainfall a year. Water is not readily available here; John doesn't irrigate, and his wells have to be dug 150 to 200 feet before they reach an aquifer.

Because of his conservation efforts, John has been able to produce a rare crop for this arid region. "We grow dryland corn, which is unheard of here, and we have been state yield winners for four years," he says. "Our soils are deep and the moisture is retained. The standard yield for corn is 220 bushels an acre under irrigation; we are raising 150 to 160 bushels with no irrigation, and we are doing it with water we save from our direct-seed system."

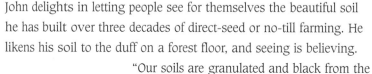

John delights in letting people see for themselves the beautiful soil he has built over three decades of direct-seed or no-till farming. He likens his soil to the duff on a forest floor, and seeing is believing. "Our soils are granulated and black from the earthworms and the microbiology," he explains. "You can dig in it with your fingers. Tillage greatly reduces earthworms because it disturbs their habitat."

John and his son Cory grow wheat, barley, peas, lentils, canola, and corn on hilly, steep ground that John's grandfather farmed. On the hillsides above the Aeschliman's land, where their neighbors conventionally farm, the contrast is dramatic. Conventional wheat farming relies on "summer fallow" which means that soil is left bare and regularly tilled with a plow to knock back the weeds and prepare it for fall planting. From John's perspective, finely tilled bare soil is anathema to soil health, and leaves the soil prone to erosion from heavy rains. It is hard not to agree with him once you see the dark rivulets of mud from his neighbor's fields that have washed downhill in deep channels onto John's stubble-covered fields. "It's just sad. They lose their water, it destroys the field, and takes it down to hard dirt," he says.

CONSERVATION

37

John tells a story about a piece of farmland he leased from a retired farmer over 30 years ago, when he was first experimenting with no-till farming. One day he was out replacing fence posts that had been buried over time, and the owner, then 85 years old, stopped by and told him that there were two fence posts below each one that he was replacing. "That was how much mud had washed down in his lifetime," explains John, "probably six or seven feet of mud!"

John's enthusiasm for direct-seed technology is matched by his frustration with conventional farming. "Direct seeding allows you to go in and plant a crop without disturbing the ecosystem," he explains. "It's just an awesome thing and it wasn't possible before. We never had the technology or the machines, and we didn't understand so many things. We know more about the surface of the moon than we do about the first 12 inches of soil in our fields.

A healthy handful of soil contains as many microrganisms as there are people on the face of earth: six billion, and only about 20 or 30 of them are named, maybe only 15 are understood."

John continues, "As you learn to work within the system Mother Nature has established, it works better, more efficiently, quicker; it's the way it should be." He adds, "Grandfather would have done it in a heartbeat, but this stuff wasn't around. But now there is no reason to continue this old system that is so highly destructive." John acknowledges that agriculture is a management intensive business and that you have to be tough to survive in it. He contends, "If you are tough, it can be hard for new information to sink in. We plow because our grandfathers did it that way, our fathers did it that way, and that's the way we do it!"

John is not totally unconventional with his farming. He takes soil tests and before he plants a new crop into the stubble of an old crop, he sprays glyphosate or Round Up to kill the plants that would compete with the new crop. For John the use of chemical inputs is an economic matter. "We could stop fertilizing now, but yields would be less," he says. "Hey, give me six or seven dollars per bushel for my wheat, and I'll cut back on farm chemicals. I would be able to do a lot of things differently, but I can't make it profitable with my costs: combines are expensive, diesel is at two dollars a gallon, labor costs are high. This is what the consumer needs to understand: it's very hard to stay in business if our costs continue to rise yet what we sell goes down in price or stays the same."

There are some promising economic incentives being developed for direct-seed farming. John is president of the Pacific Northwest Direct Seed Association, an organization that has sold carbon credits to energy producers. Direct seeding keeps the carbon dioxide that plants have sequestered from the air in the soil, whereas tilling

accelerates the decomposition of plant material, releasing more carbon dioxide into the atmosphere. John also sells some of his wheat to a new regional company that is marketing the benefits of supporting local, no-till farming.

"When you no-till, your fields are covered with residue," says John, pointing to a field of stubble. "To me that looks great, but to a farmer that does conventional farming, it is ugly. But that straw protects the soil from erosion and provides cover for the birds." John sees lots of wildlife on his land—deer, three or four species of owls, hawks, badgers, porcupines, pheasants, chuckers, quail, and coyotes. He reports, "We have some land in the Conservation Reserve Program, and it's full of wildlife. People that hunt here say this is the best bird hunting in the country, and that is because of the direct seeding that we do. The birds never get disrupted, they just move out of the way when the tractor goes past."

John knows that making the switch to direct seeding is a daunting challenge for many farmers. "It has to be your calling," he says. "You have to have the tenacity. People say it's too hard, it's too much work, and it won't pencil out. For that guy it will never work because he has already made up his mind. For the guy who wants to look at it, work on it, and figure it out, it will work."

John adds, "People always ask, 'what do your neighbors think?' I usually reply, 'Oh, some probably think I fell off the last turnip truck that went through town and bumped my head!' But you know what, they are doing now what we were doing 20 years ago. It's becoming acceptable. You soon learn that if you are an innovator you are usually 20 or 30 years ahead of the pack." He continues, "Thomas Edison went broke three times before he invented the light bulb. It doesn't always work the first time, you've got to stick with it. With the Lord's help providing the rain and the natural systems to make it all work, perseverance will conquer all."

LEONETTI CELLAR

Winemaking has been a tradition in the Leonetti family since Francesco Leonetti immigrated to the Walla Walla Valley from Italy in 1905. Francesco was a truck gardener who grew an acre of grapes on the side to make wine for his family's consumption. As a child, Gary Figgins sipped diluted wine served by his grandfather, Francesco. In 1974, Gary planted his first acre of grapes above his grandparent's original homestead. Gary taught himself winemaking and produced his first wines in his basement under the Leonetti name in 1978. From the start his wines were in demand, and he soon quit his job as a machinist to pursue winemaking full time. Today, Leonetti wines are available only in upscale restaurants or through the company's mailing list.

Chris Figgins is the next generation of winemakers in the Leonetti family, and he is grafting new ideas about vineyard management and resource conservation onto the family business. Chris collaborates with his father Gary in winemaking and is the lead viticulturist, overseeing management of nearly 200 acres of vineyards in four different locations in the Walla Walla Valley. Chris has a degree in horticulture, but feels there were some essential gaps in his formal education. "We spent so much time on soil structure and chemical makeup and very little time on soil biology," he says.

A presentation on the soil food web opened Chris's eyes to the relationship between soil biology, overall plant health, and fruit quality. He remembers, "The next spring we ran soil biology assays, and not surprisingly given what I know now, the vineyards with the highest mycorrhizal colonization had the best soil quality and health and produced the highest aromatics and the highest quality of wine." That data convinced Chris to move toward more biology-centered soil management in the fields.

With his new interest in fostering the symbiotic relationship between beneficial fungus in the soil and the roots of his grape plants, Chris began applying compost teas to the vineyards to enhance natural nutrient cycling. Chris reports, "We have seen truly amazing results. Mycorrhizal colonization went from zero to 32 percent in one year using compost tea, and fungal counts and bacterial biomass went up dramatically."

There are also more visible differences that distinguish Leonetti's vineyards from others in the region. For one thing, grasses and small flowering plants are allowed to grow on the rows between the grapes. "The wine industry is so visual in terms of the consumer and the media, that you just want everything looking perfect, like a golf course," says Chris, "when that is not necessarily the best thing for the soil." Chris points out a delicate purple vetch tucked under the grape vines and notes, "Having a pollen source is great; beneficial insects require pollen to feed on when there are no pests available. We tolerate pests as long as they don't get out of control. The key is to have things balanced."

Reducing chemical use has been a goal for Chris for quite some time. The only pesticide spray they have used in the past three years was a fungicide spray for powdery mildew. And where they once used two to three under row herbicides a year, now they use just one in the early spring, and otherwise rely on mechanical and hand weeding.

Other techniques Chris is employing to improve soil biology involve reducing bare soil. He explains, "We have recently bought a side discharge mower to blow grass into the berm row to get organic matter there." When there is enough rainfall, Chris will plant grass or clover as a cover crop between the rows. Chris has also planted lavender and wild roses in places that are not in grape production. He explains, "We use to spray fence rows to keep the weeds down. Why not plant a beneficial plant there that you don't have to spray? It covers the ground and chokes

the weeds out; you get a benefit from it and you lower your chemical and labor costs. It's a no-brainer, but until you do it and it works, you really don't think about it." Now when Chris plants a vineyard, he intentionally leaves an area out of production to foster biological diversity. "With a monoculture you have much wilder swings in your pest populations," he reports.

Drip irrigation is another tool that Chris feels is well worth the cost, particularly in this dry region. "It is the only way to go out here," he explains. "It allows us to conserve a lot of water and to have complete control. We can really shut down vines when we want to, and then we can spoon feed them until harvest which increases intensity and flavors in the grapes without growing big berries."

Chris has noticed a marked improvement in the soil's ability to absorb water since he has focused on building soil. He reports, "When we first started farming this piece of ground that had been farmed conventionally, we couldn't run drip irrigation for 12 hours. The ground couldn't absorb the water, it would just start running off. Since we have gotten the organic matter up, we can now run water for 24 to 36 hours—a vast improvement."

Chris's interest in conservation starts with the soil, but it doesn't end there. Leonetti Cellar recently moved away from styrofoam packaging to recycled cardboard. After a few years of use, their oak barrels are cut in half and sold as planters. Chris is eyeing some of the vineyard's ridges in this windy valley for the possibility of locating energy-generating turbines. And an on-site compost operation is in the works that could accept crushed skin and seeds from the thirty-some winemakers that have now joined Leonetti in the Walla Walla Valley.

Chris has been active with a group of local growers creating local sustainable viticulture standards for Walla Walla Valley, under the name Vinea. Explains Chris, "We have always wanted to be green and light, but we never wanted to push organics because we don't have that marketing need and we don't want to be pigeonholed into that." Under the Vinea program, vineyards can only use chemicals with a maximum half-life of 90 days. Chris says, "I think if we can set an example—have it be economically feasible and have quality make a leap forward—there is no way people won't get on board. Furthermore, the Vinea program employs a framework which recognizes that every vineyard has different issues and challenges: soils, rain, cover cropping capabilities, neighbors. It is not a one-size fits all program."

The growth of the wine industry in the Walla Walla Valley has been staggering since Leonetti Cellar opened the valley's first winery. "Wineries are changing the face of agriculture in this valley," says Chris. "It used to be 'wheat is king' around here. Now, wine is becoming the premier crop, along with sweet onions. It's neat to have the diversity and the value added; that is what it is going to take in the global economy."

Chris notes that while he is the one bringing in the changes, his dad is all for it. He says, "My dad has always been on the green side of things. We are in a unique position compared to other kinds of agriculture because with vertical integration we can pass on any added costs." In the long term, Chris doesn't think farming sustainably will be more expensive, in fact, he thinks it will be less so. He concludes, "It falls in so well with how we operate. We are second generation, full-time winegrowers. Everything we do is not for us or for quick profit: it is to be long-term, stable, and generational—in our business, in our soils, in everything."

LIMBERLOST TREE FARM

Herb and Grace Payne are retirees with a passion for the land that has been handed down in their family for close to a century. Herb was a stockbroker and Grace was a teacher during their professional years, but now the Paynes are spending their retirement planting and caring for the 130-acre tree farm in Skagit County they call Limberlost.

Herb's grandfather bought the first piece of this parcel in 1914 for one dollar an acre. "This property is just loaded with springs," says Herb, "that's why my grandfather bought it, for the water and the swamp." Herb's grandfather hauled logs across the property for the English Logging Company, and Herb has pictures from that time showing ten men sitting on a huge log that fills a railroad car. The stumps from those harvests now fuel the Paynes' long-term ambitions for this property: they want cedars of that size to stand here again—a goal that will take several generations to achieve.

It hasn't been easy to keep the parcel intact over the years, and Herb's family hasn't always agreed on how to manage it. "My grandfather's Irish and the Irish divided all their property up so there's nothing left for anybody— specially if you have ten kids,"

he laughs. Herb's grandfather died in 1950 and in 1952, during the Korean War, his grandmother had the property logged. When Herb's parents died, the property passed to the two remaining sons and the family of a deceased son. Because the new owners did not agree on how the land should be managed, the future of the forest on the property was uncertain. Fortunately, Herb was able to buy out the other family members in 2004, and he and Grace assumed ownership of the full 130 acres.

Today Herb and Grace only harvest alders and other hardwoods, and are replanting and nurturing cedars, in the hope of re-creating the old-growth stands that were here a century ago. If the topic is not family history, Herb lets Grace do most of the talking because he says she is better at sticking to the point. Grace summarizes their land management objectives: "When we inherited this land, our goal was the same as it is anyplace we go—whether it's a house or a campsite or a beach—to make it better than it was before we came."

Herb and Grace realized that the plantation on the upper portion of the land was being taken over by alder; they trimmed out a lot of alder so that the fir would grow. Grace says, "Cedar is what is suited

CONSERVATION

for this property because it's on the north side, it's wet, and it's shaded. We want to bring it up to capacity production and pass it on to our children with the understanding that it will be kept as a tree farm. Our goal is to leave the conifers; we want the firs that are here to grow as big as the stumps."

In order to meet their goals, the Paynes found there were some significant legal hurdles to overcome. Their property is saturated by springs and streams, including one that meets the criteria as fish-bearing, even though a manmade barrier downstream has meant that no fish have actually been there for almost a century. State forestry regulations require 150-foot buffers on either side of any potential fish-bearing stream, which is a serious limitation in the Paynes' case, given their relatively small property size. However, the same regulations also allow for small forest landowners to apply for an exception to this rule if their long-term management goals meet or exceed stream protection standards. It means a daunting array of government paperwork and forest planning, but the Paynes persevered, with some assistance from their county Conservation District and the Washington Department of Natural Resources.

Says Herb, "We personally hugged 583 trees along the stream to get their diameters, along with species and tree type." This information was used to determine the desired future condition of the land, which was an important factor in the state's assessment of the proposal. Herb recalls, "It required eight to ten people from various agencies and locations around the state to gather at our property to walk the proposed area along the stream, take notes, and come to a consensus regarding our proposal." The consensus was in the Paynes' favor, because their plan to selectively log only deciduous trees and replant with cedars will, over the long-term, better serve forest and stream health. The Paynes agreed to leave a 50-foot stream buffer untouched and harvest only during the dry season to protect the fragile wetland soils.

In 2001, the Paynes hired a forester to log about four acres of alder; the next year they planted 1000 cedars on that land. In each of the subsequent years, they have logged a similar-sized parcel and planted 1000 trees to fill in the logged area. Even though cedars are suited to the conditions on this property, they are not necessarily easy to grow when you are trying to accelerate the process of forest succession. It takes five to seven years of cutting back brush before the cedars are safely established. That is why the Paynes limit their harvest and replanting to five-acre units. Says Grace, "We can only handle five acres physically because our family plants the trees ourselves. And after a planting there is so much work in keeping the brush down and thinning around them."

Herb adds, "What we haven't done is spray the brush. The elderberry and shrubs are eaten by the deer; they nibble it and nibble it. The big companies have trouble on their cedar plantations because they kill out all the other foliage, so all the deer have to eat is the cedar, but we are not bothered by the deer." Herb uses the old stumps as guideposts for the future forest. "When you are in a swampy area, you plant where the old stumps are. We'll plant four seedlings in a circle around a stump because that is the only place the conifers have grown."

Herb and Grace earn enough money from their small selective harvests to pay their property taxes. "That's about it – it helped us to buy 30 acres of my brother's that almost got away from the family," says Herb. "Keep it in the family, that's the key so that it doesn't get subdivided." Herb and Grace are planning for the long-term continuity of their forest management plan, and part of that is planning harvests that can at least pay for property and inheritance taxes that will come up. Says Grace, "We don't know what is ahead of us. Our kids might have enough money when we die that it wouldn't be a factor, but we don't know. If we leave them some trees, that seems more lasting than money. They could do the same thing that we are doing: take five acres at a time and then pay the taxes so they wouldn't have to give up this property."

One of the bigger challenges in implementing stream restoration on an effective scale is coordinating the efforts of the various landowners that impact a particular waterway. Serendipitously, just downstream from the Paynes' parcel where Bulson Creek passes under state highway 534, the state's Department of Transportation installed a new culvert, restoring fish passage. A bit further on, a retired farmer and his wife chose to fence out their cows, plant trees, and put an easement on their quarter-mile of Bulson Creek frontage.

Herb remembers his father fishing in Bulson Creek, and he is optimistic about the prospects for fish with the recent stream improvements. "Cutthroat should be in the stream, and I was told there was a run of summer steelhead. They found some silver redds down below between the hill ditch and that big new culvert. I wouldn't be surprised if there might not be some chum come up!" With the work and plans the Paynes have in place, Herb's optimism seems well founded.

MOCCASIN LAKE RANCH

The Methow watershed in north central Washington spans an area larger than Rhode Island. The terrain of the scenic valley ranges from the snowy peaks of the North Cascade Mountains to dry Columbia River plateau. Most visitors and residents stay close to the valley floor where the sense of being surrounded by expansive open hillsides is heightened by the year-round sight of the snow capped peaks of the Pasaytan Wilderness. Though the area is a very popular destination for recreational users and new residents, local folks are working to assure that some landscapes don't change.

Moccasin Lake Ranch sprawls over 2200 acres and includes several small mountains and its own lake. Three years ago, the Pigott and Beatty families decided to protect more than half of their property from development, forever. A conservation easement held by the Methow Conservancy assures that all of their land above 2100 feet will remain as it is now—wide open meadows fringed in ponderosa and lodgepole pines. The 1400-acre Moccasin Lake Ranch easement is the most visible of the Methow Conservancy's conservation easements; it comprises much of the skyline visible from Highway 20 south of Winthrop.

Jim Pigott explains, "There won't be development or any building on the hillsides. We can continue doing what we have been doing agriculturally—putting in new fencing, water troughs, and barns. Those are all more of the same, so that's okay, but not buildings—residential or otherwise."

The lowlands of Moccasin Lake Ranch are leased for alfalfa production. On the lower cultivated ground, the ranch is moving away from the use of pesticides. "We are on the way to becoming an organic grower," says Jim. "It takes about five years of no pesticides and we are about three years into it."

Although Moccasin Lake is still a working cattle ranch, much has been done to assure that the land provides for more species than cattle, including an abundance of birds and wildlife. Unique upland aspen groves are being irrigated and fenced to protect them from cattle. The groves provide cover for a variety of wildlife and forage for a thriving deer population. To conserve water, over three miles of irrigation pipe have been buried and two ponds have been lined.

The perimeter of Moccasin Lake is fenced to limit cattle access to two small locations. Jim explains, "It's a green, lush area that provides a lot of habitat for birds and small animals as well as natural food for trout." The ranch runs a rainbow trout fishing program that is open to the public on a fee basis in the spring and fall. Jim says, "It's all catch and release, barbless-hook fly fishing and a guide is always in attendance to keep the fishing in control." The ranch allows no more than six people out on the lake at one time, and only permits fishing two or three days each week.

Since 1997, Moccasin Lake Ranch has been home for Methow Valley Riding Unlimited, an equestrian organization that specializes in working with riders of all ages who need special assistance. Jim explains, "We built a riding ring primarily to use for therapeutic riding

CONSERVATION

lessons, but we also use it for general horsemanship activities including private lessons in dressage and vaulting." Developing an appropriate riding arena required serious engineering. It was important to get the arena absolutely level so that it could accommodate riders of various skill levels, some of whom arrive at the ranch in wheelchairs. A wheelchair ramp provides riders the opportunity to bridle and curry their own horses. "That activity in itself has therapeutic value," says Jim. "We are pleased to be in a position to give all of these riders access to specially trained horses and staff."

Jim's office at the ranch contains evidence of the undomesticated animals that also make this land their home. He is quick to point out, "Every one of the animals on display here have a nuisance story—it wasn't just a matter of going out and shooting them for target practice." The bobcat that looks down from the rafters was caught in the chicken coop after a hearty dinner; the cougar standing next to a coyote helped himself to a neighbor's dog. Such

nuisance animals are either trapped and removed or occasionally become exhibits at the ranch office.

Jim and his family have also worked to enhance the historical value of their land. There is a homestead cabin built around 1908 in a remote part of the ranch that they are rebuilding for preservation. This has proved to be no small job as the cabin had to be lifted off its original stone footings so that a proper foundation could be installed.

"Preservation, conservation and production all blend together on Moccasin Lake Ranch," says Jim. On the ranch's uplands, the cattle are widely dispersed across brown brushy hills where grass grows as high as their bellies. Not every rancher has a professional wildlife biologist on retainer, but Jim and his family have more diverse goals for this land than just raising beef cattle. Their work and commitment ensures that on Moccasin Lake Ranch there will be room for other species and recreational activities to prosper well into the future.

READ, DEANNA & JEREMY SMITH FAMILY FARMS

Read Smith is a farmer, rancher, and soil conservation advocate who practices what he preaches on his 8000-acre ranch in the grassland prairies of eastern Washington's Palouse region. "I am a sixth generation farmer, the third generation of my family to live on this ground," he says. Read and his youngest son, Jeremy, are carrying on the mostly dryland farming that Read's grandfather started here in the 1930s. They raise hard red and white wheat, soft white wheat, barley, peas, lentils, oats, canola, mustard, safflower, sunflower, millet, and alfalfa hay, and run some cows.

Read considers himself fortunate because both his grandfather and father were good stewards of the land. He says, "They used all the technology that was available back during their times, and as a result I have a wonderful resource here." Read has continued this tradition. In 1976, he began experimenting with direct-seed systems; by 1997 he had converted the entire farm to direct-seed.

Direct seeding means that Read never plows his fields, instead using a machine called a drill to punch seeds into the stubble of last season's crops. Where his conventional neighbors have tidy furrows carved into the soil, Read leaves no ground bare. He explains, "In conventionally tilled soil there is a fraction of the living organisms in

the ground because that system effectively minimizes the amount of worms, microbes, and wonderful friendly bugs that are present by the millions in healthy soils." In addition to healthier soils, Read's fuel costs have gone down and direct seeding saves him time.

Where some farmers cultivate every inch of their acreage, Read plants a buffer of native grasses, legumes, trees, and shrubs around every field he plants. He has put a full quarter of his property—the less productive cropland—into the Conservation Reserve Program. "It is a pretty good trade off because you are in effect renting that land to the government for a period of time, and you keep it in a sustainable grassland system that is beneficial to wildlife and the environment," he explains. "If there is a way my descendants can cultivate these very steep hillsides efficiently and in an environmentally friendly way, that's great, these topsoils will still be here. I am just trying to protect the resource now so there will be that option in the future."

From the high points on Read's property, the golden grasslands roll on unhindered out toward the horizon. It is an arid land, averaging less than 15 inches of rain a year, but water has played a major role in shaping the landscape. The remains of ice age floods are still evident here. Says Read, "Often I get up on these hills and think about when Lake Missoula broke loose and that water was going by here at 60 to 70 miles per hour. What a roar that must have made."

Smaller-scale mudflows are a major concern for Read today. He reports, "The topsoil here in the Palouse is 40 percent gone today. We have only been farming for four generations and we have lost ten percent of the topsoil with every generation." He continues, "Technology has masked the loss of productive capacity of these soils and as a result people are ignoring the loss of that resource." Not only is the topsoil eroding, but the subsoils are washing off the hills and covering the good bottomland soils. Read feels strongly that moving away from tillage is a key to remedying this problem.

Read has worked diligently to change the ways of conventional farmers. He has been a Conservation District official since 1974, including a stint as the president of the National Association of Conservation Districts. The benefits of direct seeding take some years to realize, and many farmers aren't willing or able to make that investment. Read says, "Its frustrating to me as a proponent of direct seed to have producers make a one or two year commitment to try direct seeding, on perhaps their most marginal ground, and then declare the system inadequate. It's not a fair test."

Read is working on all fronts to be a good steward of his resources. He carefully cultivates wildlife habitat and ponds as part of his ranch operation. "It's very rewarding to see wildlife respond to things that you can do," he says. The ranch supports whitetail and mule deer, upland game birds, migratory waterfowl, songbirds, birds of prey, and other predators. Read prefers wildlife watching to hunting himself, but he is practical about the role of hunting. He permits fee hunting on his land to a small number of people that agree to a long list of requirements. All proceeds from this hunting are invested in further improvements to wildlife habitat.

To develop a system for rotational grazing, Read needed to find a way to cross fence his pastures and still have access to water. He was able to install a solar-powered pump in an abandoned wellhead, allowing him to divide one large pasture into four small sections around a centralized trough. He explains, "It will be easy to rotate cattle among these four pastures, and there will be more efficient use of the range because they will eat the less desirable crops, facilitating proper rotational grazing."

Read has been pleased with the reliability of his small investment in solar energy and is enthusiastic about the role farmers and ranchers may play nationally in renewable energy production. "The greatest potential that the Pacific Northwest and all of rural America has is green energy production," he says. "Rather than raising crops for less than the cost of production and shipping them abroad, why don't we keep those products at home producing power." Read believes that using crop residues and other agricultural sources for biofuels is an area that is only now being developed and researched to its full potential.

While the economics of wheat farming are challenging in today's markets, Read is confident that this can change with U.S. consumers demanding more accountability and security in their food production, and with the possibilities offered by green energy. "I am more optimistic today than I was 20 years ago about the potential of agriculture and rural America surviving," claims Read. "It's going to be a different world; it's not going to be like what we see now." Read does not share the doom and gloom outlook that he finds in rural coffee shops. He says, "I see great opportunity around the corner. I just hope we can survive the transition between now and then, and keep as many people on the farms as possible." He concludes, "When we get to the point where we are attracting young people back to agriculture because it's a good alternative to the city lifestyle, then we will have accomplished that goal."

CONSERVATION

TWO RIVERS FARM

At the confluence of Icicle Creek and the Wenatchee River, just a few miles east of the Bavarian-themed town of Leavenworth, you might see salmon spawning, boaters picnicking on a wide sandy beach, or bear sloshing across to an island preserve. What you may not notice is that this scenic piece of land is also in agricultural production. Here Nancy Denson and Nick Stemm grow gourmet organic produce on a farm that values what grows naturally as much as what is cultivated.

Nancy and Nick gave up life in the big city two years ago to pursue a path they once rejected. Nancy explains, "Nick was raised in this valley on a commercial apple orchard. His parents wanted us to take over the orchard in our twenties, but we chose to go live in Seattle. When we reached a certain age, we thought, 'We wish we would have kept the farm.'" By then they couldn't go back to the farm Nick grew up on, but that didn't stop them.

Nancy and Nick purchased a 26 acre narrow peninsula of land that stretches out between Icicle Creek and the Wenatchee River and named it, appropriately, Two Rivers Farm. The parcel includes a mile of riverfront property and offers spectacular views. Nancy explains, "In October there is a fall run of chinook that come up the Wenatchee River and their spawning ground is right here! We can sit in the house eating breakfast and watch the salmon spawn. Then in the spring, they come upriver on the other side, on Icicle Creek." Protecting these salmon populations and the other wildlife that share their land have driven a lot of Nancy and Nick's decisions on how to manage their farm.

Nancy and Nick decided to run the farm using organic methods because of the rivers and the fragile natural environment. They grow a wide variety of fruits and vegetables on the inland portion of their property. They also raise a handful of cattle and sheep, about 40 heirloom turkeys, and 650 chickens for the meat as well as for the fertile manure that is the base of the on-farm composting operation. They pay extra to bed their animals on organic straw and feed them organic hay and grains so that the compost they make is approved for their certified organic program. Says Nancy, "Our certification is kind of secondary. Our customers don't really care if we're certified, but I care. It's a good way for me to make sure I am checking my methods because there is a lot of guidance in the organic rules."

Two Rivers Farm sells most of its organic produce to Visconti's, a restaurant in Leavenworth. They also have a small CSA program.

Nick, who just recently began farming full time, admits, "Our limitation is our gardening skills. We have enough marketing skills. We have to learn how to grow better in some of our areas of production."

Local customers buy organic poultry, eggs and lamb directly. "The meat all goes to private customers that buy it off the farm," says Nancy. "I send out a memo to my meat customers in the spring, and I sell out for the upcoming season usually within a week or two." Nancy knows they could sell more meat, but more animals would mean that they would have to worry about where the manure would go and if it was getting into the river. "At this level," she says, "my stocking rates are so low that I don't have to do a lot of math to figure out how much manure I can put on the land."

Nick and Nancy are mindful of the well-being of their domestic animals. Nancy explains, "We are members of the Humane Farming Association, so there are a lot of management choices we make that are based on the nature of the animals, as well as the impact on the land. Our animals are out to pasture during the grass season. In the wintertime when the snow is deep, they are in paddocks and open air sheds with deep bedding. We don't believe in barns for animals. That's how you get illness." Nick and Nancy use a rotational grazing system where they can, running "very free-ranging chickens" over next year's garden plots.

Wild animals are also considered in the management plans. Half of their land is dedicated to wildlife habitat, including a five acre island in the river. Nancy says, "We are undecided whether to keep the island as a green belt or give it to an organization such as The Nature Conservancy. We like having it there because it is a place where predators can go: they are so pressured here in the valley. The island gives the bears and the coyotes an undisturbed place to

exist. We don't want them over here with our chickens, but we are trying to live and let live with them."

Uninvited humans are even welcome on a sandy beach revealed each summer at the confluence of the Icicle and Wenatchee rivers. Nancy says, "This is a really popular beach with floaters. We don't mind people using it as long as they don't leave trash." Nancy sees the river as a resource for the valley, and likes that it's accessible to the public, no matter what income. "It's there for everybody to use," she says.

The health of this land and the rivers that border it drive farm management decisions more than market demand or aesthetics. "We have learned to ignore a lot," says Nancy walking past a wall of brush that might make another farmer's fingers twitch. "I like an open woodland, but we've gone to goats instead of bulldozers because the soil is too fragile for big machinery. There is more brush here than two goats can eat, but I don't want them to eat it all. We need wildlife habitat and the brush is part of that too." Nick adds, "All of this surrounding natural vegetation is a perfect habitat for beneficial insects."

Pointing to long rows of crisp lettuce being misted under shade cover, Nancy acknowledges, "This is the only place I have allowed a battle against nature. If you are trying to grow lettuce over here in 100 degree heat, you have to have some kind of system, and lettuce is one of our main crops for the restaurants, so we're trying to sustain that."

For the past two years a pair of bald eagles have raised their young in a tall pine on Two Rivers Farm, and while some chickens may be at risk, Nick and Nancy prefer to celebrate the coexistence of the cultivated and the wild on their farm.

FARMING AND RANCHING

Increasingly, Washington's consumers want to make informed choices when deciding how to feed themselves and their families. They want to know where their food comes from, how the land and animals have been treated, and how surrounding natural resources are impacted. Consumers are beginning to realize that how they spend their food dollars can directly affect their region's economy and environment.

Many farmers and ranchers in Washington see their work as more than just food production. They consider themselves stewards of natural resources, protectors of safe food systems, and supporters of rural traditions.

The individuals in this chapter represent the potential for Washington's farmers and ranchers to feed local residents from the landscapes that surround them, while also protecting natural resources for future generations.

BLUE HERON FARM

As a first-generation organic farmer, Anne Schwartz's story is closely linked to the evolution of the organic industry as a whole. She has grown vegetables for wholesale distribution to natural food cooperatives across the country; she has worked for a large organic farm that was ultimately bought out by General Mills; and now she is primarily direct marketing her produce through farmers' markets and two home delivery businesses. Anne seems to have the energy of several people, also running a tree nursery business and actively engaging in her community and in agricultural policy debates.

"My husband Mike and I both fled New Jersey in the mid-70s," remembers Anne. "We had friends here in the Northwest and so we wandered this way. Back then the Skagit Valley was known as the 'Magic Skagit' and for good reason. It was a welcoming, abundant, beautiful, resource-rich place to stumble into."

At the time, people from all over the United States were flocking to the rural Northwest as part of the "back to the land movement." The movement consisted mainly of college-educated kids, many of whom were active with the Vietnam War, the environmental movement, and civil rights issues. Anne recalls, "We had a naive expectation that the world was going to crash and burn, and we wanted to be as far away from that crash and burn as we possibly could be." While many of these young people had no intention of farming, for Anne it was a perfect fit. She explains, "The more time I spent farming, the more I realized that it fit my personal need for being outside, as well as being mentally stimulated. Coming to farming as a suburban raised person, I quickly realized that this was an area I could learn about forever; there was no lack of challenges."

Anne and Mike settled in the upper Skagit Valley, in the foothills of the Cascades, surrounded on all sides by the Mount Baker Snoqualmie National Forest. Anne remembers, "The dairy farmer I worked for back in the '70s used to say this country is good for growing two things: trees and grass. We took that to heart." Thirty years later, Anne and Mike grow 170 kinds of bamboo, which is essentially a large grass. Anne says, "Three of our ten acres are in bamboo groves; it is a big part of our business."

Anne and Mike market their bamboo all over the Puget Sound and in California, and have brokers that ship it around the country. The bamboo is mostly used for privacy hedges in the Northwest and in

Asian-influenced gardens. As for Blue Heron being both a nursery and a vegetable farm, Anne likes the balance: "The blend of nursery and horticulture crops works well for us to spread out the workload, because the nursery industry really gears up in February and is done by the end of May." Once the ground starts to dry out a bit in the late spring, Anne gets vegetable transplants in the ground to get ready for the early season crops.

All the produce Anne grows is certified organic and marketed directly to consumers. She explains, "In 1982, there were twenty-eight natural food wholesale distribution companies in the United States, and in 2003 there were four. I used to grow crops that I sold through produce brokers and organic processors; as they stopped buying from me, I started looking for other markets."

Anne points out that while organic farmers have always had an interest in selling locally, as they've lost their access to wholesale markets due to consolidation, they have had to shift to new markets and increase crop diversity. She says, "With the rise in popularity of farmers' markets and community-supported agriculture (CSA), producers are changing what they grow to appeal to direct marketing opportunities. The future for farmers across the country is going to rely on more people being interested in supporting their local or regional growers."

Anne currently sells produce at four farmers' markets, from Seattle's Pike Place Market to the Methow Valley; she also grows for two Seattle-based CSAs, one that delivers directly to customers. To achieve a colorful market table, Anne grows snap and sugar peas, raspberries and blueberries, several kinds of beans, summer and winter squash, broccoli, cabbage, cauliflower, basil, sweet corn, and lettuce. She tried to do a local CSA several years ago, but found she didn't have the customer base in her remote part of the valley. Anne explains, "Rural Washington is not exactly your mecca for buying high-end organic food, but people do buy potatoes, onions, winter squash, and carrots for winter storage. I found it a nice niche growing fall and winter crops for my neighbors who like to stock up on storage vegetables."

Through involvement in local natural food cooperatives, Anne found that her passion for farming extended beyond organic growing all the way to an "undying interest" in agricultural policy. "Even way back in the '70s, I was involved with folks who were working to change our food system," she explains. "I have made a study of food and food policy largely from the perspective of rural livelihood, and the more I study, the more I am convinced that rejuvenating rural America is a critical challenge for our country."

Anne has long been on the bandwagon that we can't afford to lose any more farmers. She explains, "Even though I am president of the organic farm group in Washington (Washington Tilth Producers), my message is local, local, local. Organic is great, try to talk to the grower, push for them to adopt more biological pest management, but we need to support local farmers even if they don't farm organically." She adds, "If we continue to lose farmers at the rate we are losing them, I think our rural communities and our society in general will be much the worse."

There is one national policy initiative that Anne is especially excited about—an expanded Farmers' Market Nutrition Program that allows low-income people to receive coupons as part of the WIC (Women, Infants and Children) and Senior Nutrition programs. The coupons can only be spent at participating local farmers' markets. "It's a way to help low-income people have access to fresh food, support local growers, and get people out in their neighborhoods," she explains. As soon as Skagit Valley was able to participate, Anne reports that receipts doubled at the farmers' market from people coming with their WIC checks. "The difference it has made for all of us is just phenomenal," she says. "The markets are crowded with people that have never been to farmers' markets before and who don't know what to do with many of our fresh vegetables." She adds, "I realized I needed to bring recipes for what to do with winter squash, what to do with a Skagit sweet onion. The next step is to educate these people about what a huge difference it makes: I have local people that work for me, and every bit of that money is going to stay in this community."

Anne is proud to be part of a generation that is working to forge new visions of land and resource use. She says, "Okay, we are not going to change the world as much as we hoped, but we are changing the world. Everybody said organics was a fad that would go away, everybody laughed at us. It is not going away, and we are having an impact and decreasing the level of pesticides being applied to the planet."

Anne concludes, "As a culture we are losing our sense of community. We are losing our time to relate to people, we are losing the time to do anything for our communities, and our culture places the accumulation of stuff as a national priority. There are many of us that don't want to live like this; we are not the majority, we don't go to malls, and we actually try to buy local. Even though my days and life tend to be overbooked in a chronic way, I know that my personal sanity is dependent on living in a rural place and having a direct relationship with the natural world."

COLVIN FAMILY RANCH

Fred Colvin's family began running cattle on this prairie land in 1854 when Ignatius Colvin arrived to find grass "as high as a horse's belly." Four generations later, the Colvins are changing how they work with the land, restoring salmon habitat and carving out a unique niche that capitalizes on their greatest asset: the lush prairie grass that covers most of their 640 acres.

Where traditional ranchers concentrate on raising the cattle that ultimately pay their bills, Fred has shifted his focus from the animals to the grass they eat—from cattle marketing to cattle feeding. He offers top quality pasture for rent to cattle dealers who pay him based on how much weight their cattle gain. Fred is happy to see the responsibility for marketing cattle rest on someone else's shoulders: "I let somebody who's set up to handle marketing do that. I take the cattle and grow them up on the grass resource we have here. That's my specialty."

Fred is not even a year into what he calls "this experiment" but he is already pleased with the results. Standing in a summer green field with black, blonde, brown, and white Herefords, Short-horns, Brahmas, Limousins, and Angus, he says, "I like the idea that these cattle are here, putting on a couple hundred pounds of weight, and I like the idea that they're leaving in a few days."

Using grazing animals to maximize grass production requires diligence. Every two or three days the animals have to be moved to new pasture, giving the grasses in the previously grazed fields a chance to regenerate and build up food reserves. Fred uses portable electric fencing to keep the cattle moving across the pasture, maximizing the productivity and health of the grass. The cattle seem used to the fence-moving routine. He remarks, "Every time I show up they get fresh grass, and they like that!"

Nose pumps operated by thirsty cattle have made accessing stream water from afar convenient. "Since the idea with strip grazing is that you're always moving the animals through the field, you've got to be able to take the water with you," explains Fred, "This type of technology enables us to do that." Fred admits he was dubious about the nose pumps to begin with, but has found them to be a godsend.

The Colvin family's picturesque red farmhouse was built in 1877 and is on the National Historic Register. The driveway is lined with maples of the same vintage. "When I grew up here we used to continuously graze," recalls Fred. "Probably the biggest change is that we've gone to dividing the prairie up into small paddocks. But we still have the basics: we've got your ground, it grows grass, the

cattle eat that, grow up, put on some weight, and then they get sold." The differences are subtle but profound. Fred explains, "We'd rather just try to fit the cattle and the management to the land. It's easiest to work with nature the way nature is operating and not try to change it."

Fred has decided not to hay this year and hopes instead to extend his grazing season by as much as 45 days, minimizing the expense of buying hay for his own small herd. This ranch is unique geologically, with low rolling hills that were formed as the last glaciers were receding. "The theory is," says Fred, "that the glaciers broke into chunks of ice, and as they would melt, water would run around those chunks of ice, eroding the soil and leaving these mounds." Haying these uneven fields, strewn with rock, mud, and the occasional beaver dam, is not something he misses.

Fred is intrigued by the possibility of direct-marketing a small amount of grass-fed beef locally. "It wasn't until the last year or two that I really believed the hype about grass-fed beef," he admits. "I guess I kept reading enough until it finally started making sense." The Colvin's are not far into it yet. "We're seeing if we can have the right kind of cattle and the right kind of pasture conditions to have an acceptable product," Fred explains. His daughter is helping with marketing the grass-fed beef, and he hopes that someday one of his daughters will take over the ranch, keeping it intact and maintaining the family tradition. "It's up to our generation to figure out how to set this thing up so future generations can manage," he adds.

Much of the prairie land that once covered southwest Washington has been lost to development and encroachment by larger vegetation. Fred explains, "Native Americans would come down here every year and they would burn these prairies in the late summer and fall so that they could get at the camas bulb, which was one of their primary food sources during the wintertime. Consequently the prairies stayed wide open like this." Fred is working on projects with the Natural Resource Conservation Service to mimic burning and encourage native species through grazing.

Prairie grass isn't the only feature that sets the Colvin ranch apart. Coho salmon spawn in the mile of Scatter Creek that cuts through the land. Fred recalls, "Dad tells a story that when he was a kid, they could go down to the creek with pitchforks and throw the salmon up on the banks of the creek." Fred hasn't seen salmon like that in his lifetime, and didn't even know they were still there until his dog turned up sick from eating raw salmon. Now, in addition to careful grazing that minimizes runoff, Fred is working to improve salmon habitat in the creek, fencing the cattle away from the banks and increasing the riparian buffer zone. The ranch is home to rare native wildflower and butterfly species as well.

"If you're going to survive in this country and do natural resource management, you've got to be in tune with what the public wants," says Fred. "Society wants clean water and they want salmon in the water. I've got to figure out: how can I meet those societal goals and still continue what I'm doing on my private property? Can we figure out how to have cattle on this ranch with that creek here that has salmon in it?" He continues, "I think the salmon would do better going through a ranch than through a housing development. I don't have any studies to tell you that, but it's my gut feeling."

Fred clearly does not miss dealing with the vagaries of the commodity cattle market, where price fluctuations can mean you don't always cover your feed costs. Says Fred, "I see custom grazing as an opportunity for some of our local ranchers to reduce risk." But he is not quite ready to convert his neighbors. "To be honest with you," he laughs, "I haven't told a lot of people. But word's gotten out, and I think maybe they're thinking 'He's kind of crazy!' And that's fine, I don't mind. We'll see what happens. I don't mind talking about it if it works, and I don't mind talking about it if it doesn't work."

DOUBLE J RANCH

Sometimes you *can* go home again. When Peter Goldmark left his family's ranch in the Okanogan Valley to go away to college, he didn't think he would ever come back. When he earned an advanced degree in molecular biology and was offered a job at Harvard, he was off and running on a different path. But before he started that job, he brought his wife back to the Double J Ranch for their honeymoon, and they never left.

"There are a lot of people who spend their lives at the bench, at the desk, or at the office, so they can run away on the weekend to get a little bit of what we have here all the time," explains Peter. "So we learn to deal with the economic stresses and appreciate every moment of living out on this land."

Peter's father and mother purchased the property after World War II from the original homesteaders. He says, "My parents lived here about 20 years and I have been here 32 years. Times were tough when they started. Some days were just flat out impossible; if the wind was blowing and it was minus 20 degrees and you were trying to keep the pipes in the house from freezing and feed and water the cattle, it was an exhausting experience." Still, it

was home. Peter declares, "Once you grow up here, nothing else is really comfortable, and that is why I am here now."

Peter's parents had run a cow-calf and wheat operation. They had been gone from the ranch for ten years when he took it over. He recalls, "The place was kind of run down. It had been run by non-owners, which is often hard on an operation." There was a lot of rehabilitation to do, especially in terms of soil conservation.

Since Peter's takeover, the cattle operation has evolved into a grass finished yearling operation. He explains, "We are working on developing local and distant markets for grass-fed cattle. We feel very strongly about the nutritional and sustainable values of raising animals in this way. They are free-ranging on native grass, and when they are sacrificed it is done in a very respectful manner." While Double J Ranch sells mostly to local consumers, they are developing a website to direct market their beef across the country.

Thirty miles of fencing and cross-fencing mean that Peter rotates cattle over the ground, never grazing heavily in one spot. He

FARMING AND RANCHING

explains, "We've added pasture ground so that we can avoid the trap that many other landowners fall into during a dry year when the grass resource suffers from overstocking."

Peter is also interested in sustaining the native grass species that have thrived in this particular climate over generations. "Native grasses are really important, they are so well adapted to this arid semi-desert ecosystem," he says. Peter's father chose this location because it was one of the last areas of native bunchgrass in the Pacific Northwest. Because he was careful not to overgraze, the stands have been maintained. Peter explains, "The great value of bunchgrass is stability and ability to cover the ground reliably even during dry periods. When you take out the bunchgrass, you get invasive species. I think those bunch grass stands are beyond valuable, so I take great care of them. Even in a drought year, I'll only use it lightly, and if necessary, I'll cut down my herd size to fit the season."

The 8000-acre ranch is almost evenly divided between pasture and cultivated ground. But while cattle prices are setting new highs, wheat is the same price it was in the early 1970s. The main change Peter has implemented in the wheat operation is minimizing the amount of summer fallow, because summer fallow is when the land is most susceptible to erosion. He explains, "This was almost always summer fallow–winter wheat country, but having half of the land exposed to wind and water erosion was unacceptable to me, so I added two crops of spring wheat into that rotation. It gives me three crops in four years instead of two." It also keeps the land covered in stubble an additional year, so it is fallow only one year out of four instead of two. Cross slope farming and water control structures are also employed to conserve the soil.

Wildlife also has a place at Double J Ranch. "We've got things that fly, things that crawl, and things that run, and we enjoy them all," Peter says. "There are a lot of hawks and eagles here that do a lot of work for us and we appreciate that. Rodents can be a big problem when you have a mixed operation of pasture and field, so from a pragmatic standpoint, predators of rodents work in our favor." Double J Ranch has 160 acres that have been kept as wildlife preserves for 30 years and these areas provide habitat for bear, deer, game birds, and many other species.

In the late 1980s Peter had the opportunity to go back into research at Washington State University. He began researching the

genetics of wild plant species to understand how seeds were able to remain dormant, with a long-term view of understanding seed dormancy as a possible weed control opportunity. In 1994, Peter decided to get into wheat breeding. He says, "Today we are on the cusp of releasing the next generation of wheat varieties—wheat that is winter hard, and resistant to foot rot and snow mold." Peter welcomed the opportunity to integrate science into his life again. He says, "It has been really rewarding to have the opportunity to bring both parts of my life together: the scientific part and the agricultural part. Through the wheat-breeding program, I am actually able to do something that has application

not just here, but in other parts of the state. I finally found a way to do science here on the ranch."

Peter's work in wheat breeding draws on his farming experience as much as his scientific background. "I think the talent is in selecting plants and stands that are going to be productive, resilient, and successful," he explains. "I think I bring a rather unique blend of some understanding and talent in the science arena and a tremendous wealth of experience. Ever since the age of ten I have been riding a wheat combine every summer, and that brings innate knowledge of the wheat plant and what is successful and what is not successful."

In addition to farming, ranching and wheat breeding, Peter serves on a handful of statewide boards, and even did a stint as the director of the Washington Department of Agriculture. He says, "I am enthusiastic about the opportunity of finding ways to put value on products other than just what the mass market will provide. I don't want to be a producer just for the lowest cost; I don't think there is any future in that for agriculture or anybody else."

"There are better things in life than money," Peter insists, "and leaving the resource better than you found it is one thing we all should strive to do. What is unusual about this region is that it is so young: the homesteaders only came here in 1918. My heavens, we haven't even been here 100 years! So we have the opportunity to treat it right."

FULL CIRCLE FARM

The weathered red barn and old tractors moving slowly through Full Circle Farm seem typical, but on closer inspection you might notice that many of the folks cleaning produce, loading boxes, or driving tractors stacked with flats of lettuce seedlings look more like college students or high tech entrepreneurs than seasoned farmers. On this land, the ideals of the organic movement have been grafted with the sophistication and market savvy of the twenty-first century, and the results have been fruitful.

Full Circle Farm has 120 acres of land in production, most of it along the Snoqualmie River, 30 miles east of Seattle. This large scale, coupled with cooperative links to other west coast organic farmers, has allowed the farm to offer greater flexibility than other subscription farms, bringing the idea of community supported agriculture (CSA) to new audiences by providing unparalleled selection, convenience, and service.

Where most CSAs provide their subscribers with a box of whatever is ripe every week from spring through fall, Full Circle Farm collaborates with other farmers to provide a wide variety of organic fruits and vegetables every week of the year. Customers can make online substitutions and opt out of some weeks to accommodate their schedules. Instead of one-size-fits-all quantity and pick-up options, Seattle area clients can choose from three box sizes and 25 pickup locations. "We are really making it as consumer-friendly and convenient as possible, and the feedback we've been getting is just amazing," says owner Andrew Stout. "People are so excited about being part of this, about the opportunity to work with the farm,

while also having it work for their lifestyle. It's 2004 and people have busy lives; they don't want to be tied to a traditional CSA."

CSA membership has recently doubled to 600 subscribers, and Andrew expects that number to double again as they add eggs, honey, cheese, bread, and shade-grown coffee to their list of offerings. Full Circle also sells to over 50 restaurants, 15 grocery stores, 12 farmers' markets, and 4 wholesalers. All of this adds up to expected annual sales of over $1.2 million.

The seed of Full Circle Farm sprouted just over a decade ago in the Midwest. Andrew had completed an internship on an organic farm in Minnesota, and was looking to start his own farm in the Northwest in partnership with his wife, Wendy Munroe, and childhood friend, John Huschle. They raised capital for the new venture by unconventional means, making and selling over 1000 egg rolls at Grateful Dead shows. Before long, they had leased five acres in North Bend. Andrew recalls, "It had just three acres of tillable ground on a beautiful mountain side, but very rocky, rough conditions. We started with a rototiller and an idea, and only novice farming skills."

Still, by the end of their first summer, they had put together 20 sample boxes with a price list, farm description, and a business card, which they drove around on a Friday to give to restaurants and grocery stores in the Seattle area. Andrew explains, "We called people back on Monday, and we started with $1600 in sales that next week, and it's climbed ever since." Every year they were able

FARMING AND RANCHING

to put a bit more acreage into production in different locations. Andrew attributes their success in those first years to their high-quality produce, and to stepping into the market at the right time.

By 2000, they were farming four different sites, and it was becoming logistically challenging to move equipment efficiently from place to place. When the roof blew off the donated trailer they were using as an office in North Bend, they decided to consolidate elsewhere. Through a service offered by King and Snohomish Counties called FarmLink that connects aspiring farmers with land and services, they were able to find an 80-acre dairy in nearby Carnation. In 2003, they purchased the adjacent 60 acres, allowing them to have more crop rotations and year-round production.

Quite a bit of work was required before they could move the whole operation to Carnation. It included pouring two hundred yards of concrete; completely power washing 80 years of debris out of barns; getting rid of all of the manure, stables, and stanchions; putting up the greenhouses; and building the infrastructure (kitchen, break rooms, office) that the place required to be operational. They did all of that while farming 40 acres. "Fortunately," Andrew recalls, "that was the year I was getting married, and we wanted to get married on the farm. There's nothing like a wedding to make a farm look good!"

Now Andrew and Wendy are reaping the benefits of their hard work. "We have a good marketing plan and a very diverse cropping mix. If one crop fails, we've got ten more to take its place." Andrew smiles and demonstrates his unflappable sales skill, "If you don't want my apples, how about my kale, my potatoes, my squash? You find they can't say no to everything. If one account is not buying much that week, we've got 15 other outlets to sell to." Andrew admits it was hard to build up to this point, but now it's able to run itself and take the stress off of them as business owners and principle farmers. Andrew still doesn't consider himself an expert farmer. "I am not a great grower," he says. "I am good, but it takes years to be great. Marketing is my strong suit—and providing service."

Full Circle Farm is also working with regulatory agencies to address environmental concerns connected to the Snoqualmie River that runs along one side of the farm, and to Griffin Creek that flows through it. It was the first farm in King County to complete a horticultural plan that addresses erosion potential, waste

management, and stream setbacks. Andrew says, "We're trying to be the exemplary farm around here and show other farmers that you can work with the government, and do things legally, and it's not a hardship, it is not something that's crippling." When the farm's irrigation ditches needed cleaning, Full Circle Farm braved the 19-month permit process that led to the ditch being sloped, graded, meandered, planted with willows, and augmented with cedar stumps. "It was not easy," Andrew points out, "but it worked, and

we were able to show the county that some parts of the process probably weren't necessary."

Partnership building clearly comes naturally to Andrew: "We're just out there saying 'yes' to whatever we can. It's exciting!" Other projects underway include: development of a compost facility that will help several local dairy farmers manage their waste; Washington State University research into better managing the flea beetle; trial cultivation of Chinese healing herbs which currently have to be imported; production of jams and salsa; and promotion of farm to cafeteria programs and institutional purchasing of local foods. "We are always trumpeting that local is better: local flavor, local fresh, local support," Andrew says.

Andrew and Wendy have themselves come full circle by offering seasonal internships to aspiring young farmers. Of last year's five interns, four have started their own farms. "We're first generation farmers," he explains, "which is unique but also becoming more common as a lot of folks from the old days have gone out of it because the food system is broken and they weren't able to survive. Now we've got new people coming in. The barriers to entry are stiff, but there is a strong desire to do well by both yourself and the community." Andrew is bullish on the future of organic farming, and his optimism and enthusiasm are contagious. "I grow food because I believe that's what I am supposed to do—grow good, healthy, organic produce for people."

GIBBS' ORGANIC PRODUCE

Grant Gibbs is a modern day pioneer who has integrated farming and forestry operations on his 80 acres into what he calls "a 1930s era fully cycling farm." In his pine-encircled valley tucked in the North Cascades, eight organic garden patches are interspersed with pasture, orchard trees, a creek, poultry and pig pens, a small scale mill, and round wood buildings he constructed from timber he selectively harvested from the steep surrounding hills.

Grant has invested almost 30 years into managing and maximizing this land's productivity, using every natural farming technique ever heard of and then some. He says, "I saw this farm as a spot to do 'permanent cultures' and pass it on generation after generation. When you plant an orchard it is not a one person lifetime thing—it goes on and on and on. The berries, the fruit trees, the forest, the riparian zone—the whole ecosystem is working together as a

permanent culture. I am not ever going to take my hayfield out of hay because I need it for the cows, and I need the cows for compost, and I need that manure for the orchard."

Grant's first lesson in farming came in the late 1960s when he decided to head for Canada instead of being drafted for the Vietnam War. "I didn't have enough money to make it to the border, so I went underground, working without a social security number as a migrant, a hobo. I rode the freights and picked orchards," he remembers. In 1975 he was able to buy a deserted dairy farm. "It was a mess," he says. "Nothing was here, no power, no wells, no fields, no road." Grant started farming organically, after having worked on chemical farms in his younger life. He explains, "I could see what it was doing to the ground and the air and the water and the people that worked it. I knew I didn't want to go down that road." But he was less certain about how he would make a living farming a different way.

Grant's goal was to make $10,000 off his land and he was happy when he first hit that goal. "My dream had come true," he says.

"Back then, who would have guessed that organic would do what it did? I thought it was going to be a problem my whole life trying to find a market, somebody who wanted to buy organic hamburger, organic pears, organic lettuce. Now the demand is such that you can basically stay home and let the phone ring and if you want to answer your phone you'll sell your whole crop."

But Grant has chosen to only sell locally, even though the demand for organic produce is far greater west of the Cascade Mountains. He refuses to haul his goods to Seattle. He says, "It is either going to get sold in this county or fed to my pigs. I'm not going to run the I-90 gamut and burn fuel. I just want to stay simple and sell what I grow within 20 miles of the farm." Even with that self-imposed limitation, demand has grown and Grant now raises produce on about two and half acres.

"Originally I had three gardens, now I have eight," he reports. "I raise six to twelve cattle depending on my hay crop—they are grass fed and people love it. My hogs and fryers are all spoken for. Customers don't have to pay me up front, but it essentially is a subscription agricultural program."

In addition to selling directly to his neighbors, Grant retails the products of his farm at a number of local farmers' markets and food co-ops. He acknowledges that the hardest work has been selling, rather than farming. "Probably the biggest challenge was accepting the fact that I was going to have to be a farmer *and* a marketer," he admits. "To farm like I do and stay in control of everything you grow, you've got to be a marketer."

Grant has designed his farming systems to assure that almost everything that comes off his land has a market or is reinvested in the fertility of the ground. He is close to meeting his goal of having no off-farm inputs. Grant keeps livestock to provide manure for fertilizing his fields. He designed and built a "pig tractor," a movable pig pen that he rotates over all eight vegetable gardens as part of his four-year rotation. He explains, "Wherever I had the pigs last summer, I plant sweet corn the next summer; after that come the leafy greens; then a tuber on the third cycle; the fourth year I do a legume before I go back to the hogs." Every second year, Grant does a light application of compost on his gardens. He makes a couple tons of chicken manure compost every year. The coarse sawdust from his Volkswagen-powered mill is used as bedding for his cows, and then as a key ingredient in the annual batch of 25 tons of cow manure compost.

Grant designed his orchard to provide an additional hay crop. He says, "I manage the orchard floor like I would the pasture because I consider it a benefit to have that long tall grass in there that offers a sanctuary for the beneficial insects." Every year he releases hundreds of dollars of beneficial insects, eight different kinds. He has been doing this for over 20 years, and monitors the populations to see if they are over-wintering and if they are keeping the pest insects in check. Grant explains, "I have had really good luck with it. Instead of a short-term 'spray the problem with a biological insecticide and call it good,' I am planning 20 years down the road on having the whole thing balanced out—the insect population will be working for me."

Taxes make up Grant's biggest yearly expense. "The best incentive that could ever happen to me would be if the county tax assessor realized that this awesome farm provides clean water, healthy forests, organic produce, organic Christmas trees, organic meats, and organic hay. If they valued that enough, they could cut my taxes a bit or even eliminate them." Grant continues, "That would be a huge help to me. I'm doing the same thing I've been doing for 30 years and everything is changing around me. All these mountain tops are getting second or third family dwellings built on them, and guess what, up go my taxes."

As the area around the farm changes, and new neighbors move in, there are a lot of things going through Grant's mind. "Maybe it's about time to bite the bullet and spend $10,000 and build a new stainless steel county approved kitchen so I can do value added food products," he wonders. "Maybe that's the way the farm can keep up with the increasing taxes and the surge of people coming into town with the big money."

In the meantime, Grant has his hands pretty full as it is. Over the past 12 years he has hosted one to five interns in a seasonal farm apprenticeship program. Now other members of his family are taking on more of the farm work. Grant's oldest son has built his home on the property and lives there with his wife, assuring continuity among the land's human inhabitants.

As he surveys the tall pines that loom around his fertile green pastures and leafy green gardens, Grant reflects that his life provides for continuous learning: "I feel like life and farming are ongoing experiments with no certainties to the outcome. Whether they fail or not, it's still a learning experience. As long as things keep changing, I'll stay on the beginning end of the learning curve."

FARMING AND RANCHING

HEDLIN FAMILY FARM

Dave Hedlin's grandfather moved from Denmark to the fertile Skagit Valley about 100 years ago and settled close to where the North Fork of the Skagit River lets out into Skagit Bay. His grandfather helped build the dikes next to the town of La Conner and bought land from the Conner family as he could afford it. Dave explains, "There are actually 17 tax statements on the 65-acre farm. Each time my granddad would make money on a seed crop or something, he would buy some land."

The dikes built by men like Dave's grandfather have kept this naturally moist land in production. Dave explains, "This is all subtidal agriculture. If this dike wasn't here, at high tide we would have about six feet of water." The wetness of the valley attracts all sorts of wildlife. Dave remarks, "Virtually everything on the Washington Department of Ecology wetland poster swims or flies across here every day." The area offers prime chinook salmon rearing habitat, with fresh river water mixed with a bit of salt water. "With the shading over the water," Dave says, "it's a nice place for salmon to equilibrate—spend a little time in that interface between fresh and salt water. It truly is a magic spot."

The Hedlin family now owns 150 acres, and farms a total of 400 acres of prime delta farmland. They have produced vegetable seed there for over 100 years—primarily cabbage seed, beet seed, and spinach seed. "If you have sauerkraut in Germany or kim chee in Korea," Dave explains, "there is a really good chance that the seed that grew that cabbage came from within ten miles of here." Pickling cucumbers, peas, wheat, bees, greenhouse tomatoes, and a farm stand are other key pieces of the operation.

Many years ago, the Hedlins set out to diversify their operation for two reasons: to help with cash flow, because cash flow on a Northwestern vegetable operation is very seasonal, and to provide more year-round employment for their people. To diversify, Dave drew on the past. He explains, "We had third generation connections with a seed company. My mother was a field rep for Lilly Seed Company—when women were not field reps. She was a remarkable woman." Hedlin Family Farm built on this link and began growing conventional and organic cauliflower and cabbage plants and providing transplanting services for the seed industry. Dave explains, "We're differentiated in the marketplace by providing these services: the seed companies don't have to own their own transplanters and they don't have to hire crews to run that equipment."

The Hedlin family also added greenhouses to expand their growing season and to insulate themselves from the risks inherent in farming in a less controlled environment. "We set out 13 years ago to figure out how to grow a good tasting greenhouse tomato," Dave says. "We worked really hard at it and I think we are about there. We grow traditional beefsteaks and about 40 different heirloom varieties." In the greenhouse operation, the only thing keeping them from organic certification is that they still use a bit of conventional fertilizer in the mix. "Organic is an emerging part of our business," explains Dave. "It makes up about 10 to15 percent of our total acreage. We are Food Alliance certified. On virtually all of our fresh market operation we just don't use any pesticides, we use some beneficial insects and that's about it."

A small, whitewashed, on-farm fresh market stand offers "you pick, we pick" berries, peppers, fresh bundles of flowers, tomatoes, and basil, all grown on the surrounding 25 acres. Dave remarks, "We started about 25 years ago growing strawberries here and evolved to this. We try to diversify our operation: I characterize it as trying to be profitable without losing track of who you are." They hit a couple farmers' markets a week and sell the rest of their produce on the farm. It has not been incredibly profitable, but Dave thinks that this sort of diversification is really important for them as a business. He adds, "The nice thing about the fruit stand and the fresh market operation is that you build a clientele of a couple hundred customers, and you have a business. If you have one customer, they own you."

Dave is downright cheerful when talking about the challenging economics of farming: "There are lots of different crops, like winter wheat, that are not particularly profitable, but they are important in our rotation to help us build up this heavy clay soil, get more organic matter, and break up disease cycles." Other pieces of Hedlin Farm's business have emerged in what Dave calls "natural fits and progressions." As an example, he notes that they run about 250 beehives for their own use because both the pickling cucumbers and the cabbage seed need bee pollination.

Dave is concerned about the issue of encroaching residential development. "Any place this nice, people see it and want to live here," Dave says. "We need lots of help to keep land in production and open space; it's an incredibly difficult job." At the Hedlins' greenhouse next to La Conner, land is worth $4000 an acre; if you walk ten more feet into La Conner, it's worth $5 to $10 a square foot. "Those are huge differentials," Dave explains, "and that puts on lots of pressure. It begs the fundamental question of why land for commercial use is worth so much more than land that produces food, open space, and wildlife habitat."

Farmers in the Skagit Valley have joined together to address the preservation of farmland, and they cooperate on more immediate farming needs as well. The Hedlins swap land with their neighbors to maintain a five-year rotation on their fields. "As the valley has tended to specialize, we have not sacrificed our rotation, but we have started trading ground more," comments Dave. "I grow a lot of pickling cucumbers, my neighbor grows a lot of potatoes. In a given year I might give him 30 acres for potatoes and take 30 acres of his for cucumbers. So you basically achieve rotation by trading ground."

Farming the same land as his ancestors has provided Dave with a unique long-term perspective. "When somebody tells me they are going to run for political office, I say, 'Try and be a hero in 30 years, not tomorrow, because you're not going to make it on tomorrow. There will always be somebody upset, but if you do the right thing with a good vision, you can be a hero in 30 years.' That's the kind of thing we need to encourage," he says, "that long-range vision."

Nash Huber & Delta Farm

The real estate ad for the Delta Farm was enticing: "Rare 100-acre farm in Sequim! This beautiful property has been surveyed into five-plus acre parcels. Outstanding views of the Olympics and neighboring farmland. Approximately 30 acres in wetlands. Lots of possibilities here."

Organic farmer Nash Huber recalls his reaction to the ad: "I was farming a piece adjacent to the Delta Farm and when I saw that land up for sale, I knew we had to act fast." Nash suggested that Seattle-based PCC Natural Markets buy the Delta Farm to preserve it as agricultural land. PCC rose to the challenge and created the

Farmland Fund to secure and preserve threatened farmland in Washington State. In July 2000, the Delta Farm became the Fund's first purchase. Now, instead of sprouting houses, a conservation easement assures that this fertile piece of the Dungeness River delta will remain in organic production forever.

Having farmed in the region for years, Nash knew firsthand how valuable the land was. "We have extremely fantastic soils," Nash enthuses. "A lot of it has to do with the mineral content of the Dungeness River water. Dungeness Valley milk has the highest butterfat content in the state. It's also the reason our carrots are so sweet." Nash's love of growing organic food and taking care of the land he farms is palpable.

"We have a microclimate here in the lower Dungeness Peninsula that is so unique. Today's a beautiful day—it's burning up in eastern Washington—and here we are with the Strait of Juan de Fuca ocean breeze blowing in." The mild climate allows Delta Farm to ship produce 12 months out of the year, alternating crops with the seasons. In the summer, they grow spinach and basil; in winter it's brussels sprouts, carrots, parsnips, and cabbage.

Nash has a background in science that informs his farming. Before he became Washington State's largest organic grower, Nash earned a degree in organic chemistry and worked as a lab technician in the Midwest, analyzing corn and soybeans. He eventually shifted over to research, which began his transformation into an organic farmer and a farmland preservation advocate. "I saw where food was headed," he says. "The grocery products division was right across the aisle from my lab, and they were making their cherry pies with red dye and modified corn starch to give it that firm gelatinous consistency—and those fake cherries."

Nash started farming on the lower Dungeness Peninsula in 1968, a little bit before the organic movement really got going. According to Nash, "At that point, the agriculture community had pretty well broken down, it had lost its spirit and its focus. We were losing our farmland; the development economy was starting to grow. It was depressing."

In 1989, legislation provided Nash his first foray into protecting farmland from development. He recalls, "I first got my teeth into it helping the county come to grips with the Growth Management Act." In fact, Nash was party to two successful lawsuits against the county to force compliance with the Act. "But when you looked at the energy it took, it seemed to me we weren't really accomplishing a whole heck of a lot," he says. "The county tried to conform to the Growth Management Act, but the big economic pressure is from the banking industry, the developers, the homebuilders association, the real estate community. You are not going to regulate that. You've got to find another answer."

Nash reverted to his real passion, hoping to set an example. "I didn't want to be perceived as somebody always suing the county," he explains. "I didn't particularly enjoy spending all my time meeting with lawyers, fussing with the county about their codes. I wanted to be farming—that's my heart, my love." To change his tack, Nash invited the public to come spend a day on his farm. "The first Farm Day Celebration was a great success," he recalls. "We had more people than we ever dreamed show up." That 1996 event that began on Nash's farm has grown to include many more farms west of the Cascades. The Western Washington Harvest Celebration Day is now observed in 12 counties and attracts some 20,000 visitors each year.

Nash was selling his organic produce through the local farmers' market when his wife received an inheritance. The couple didn't have much trouble deciding what to do with it. Nash recalls, "I said, we can leave this in stocks, or we can put it in what we believe in." They invested the money in a packing shed and the business grew from there.

A long time participant in and organizer of farmers' markets, Nash is a strong advocate of direct sales rather than commodity markets for farmers. "Not only do you need to have control of your price," he explains, "you have to have a relationship with your customer. That's why the farmers' markets are so valuable, your customer tells you what they are looking for, what to grow, and how to grow it. This can give you a year or two lead on the big commercial operations."

"For example," he recalls, "our customers at the farmers' market were saying, 'Boy you guys grow good carrots, these are the best carrots I've ever eaten.' Finally, even as thick headed as I am, I heard that." Nash contacted PCC Natural Markets, and in 1991 they agreed to try out his carrots. Other produce soon followed and a great partnership began to take root, nourished by the fertile soil of the Dungeness River Valley.

PCC is the largest retail food co-op in the nation, with 40,000 members and seven stores in the Seattle area. PCC has historically been willing to back its suppliers of local and organic food with loans. But back when Nash called in 1999 to say that Delta Farm was being marketed for development, a bolder strategy emerged. That's when the PCC Farmland Fund raised and borrowed the money needed to buy the land, and Nash now uses Delta Farm to train the next generation of organic farmers. He cultivates young farming interns who are willing to invest more than a summer. Nash says, "We've got the land, now we need the ideas and the people."

Currently, 25 to 30 people work with Nash on a 300-acre patchwork of parcels bordered by new housing in the Dungeness Valley. Along with produce, Nash raises hormone- and antibiotic-free pigs, seed for organic cover crops, and he is experimenting with poultry. Some of his acreage is devoted to providing habitat and food for non-paying

consumers, such as migratory birds and the endangered Taylor's Checkerspot butterfly.

Nash spends his days shuttling between numerous pieces of leased land. "Basically my job is to drive around and talk," he laughs. He gestures out over a green field dotted with workers: "With conservation easements on the land and what the PCC Farmland Fund did, there's a future for these folks because we've actually saved the land. You don't make enough money farming to be able to buy this land. The only way I've been able to put this farm together is by setting an example and convincing people to work with me on it."

Since its purchase of the Delta Farm, the PCC Farmland Fund has subsequently saved several other parcels of farmland in Washington, and they aren't through with the Dungeness Valley yet. In partnership with nine different state and local organizations, the Farmland Fund has helped create the Dungeness River Estuary Project. This collaborative effort aims to protect a mosaic of nearly 300 acres of farmland and wildlife habitat near the mouth of the Dungeness River, and to restore the river to its natural bed. Says Nash, "We want to see that land stay in agriculture because it's easier for a farmer to deal with a flood than a bunch of houses."

S & S Homestead Farm

Henning Sehmsdorf has stag bladders stuffed with rotting chamomile flowers hanging from the laden apple trees in his orchard. Innocuous garden stakes in the dirt indicate where similar concoctions ferment in hollow cow horns. After a year of curing, these potions will be greatly diluted in water and sprayed on the fruits and vegetables that sustain Henning, his family, and their livestock throughout the year. The methods on S & S Homestead Farm may seem outlandish, but the results are decidedly not.

Take the fact that Henning, who is approaching 70, has had virtually no health expenses, except for regular check-ups, in the thirty-plus years that he has grown his own food. Neither have his wife and kids. No cavities either. Even the livestock—cows, sheep, pigs, and chickens—who are fed homegrown and homemade feed, require no veterinary care, no medication.

"We don't have any disease," says Henning definitively. "Our pest management regime is basically healthy soils. Healthy soils grow healthy plants, grow healthy animals, grow healthy people. We do no monocropping of any kind; biodiversity attracts beneficial insects and pollinators and drives away the organisms that will be disease carriers. We don't introduce any beneficial insects either; the microorganisms in the soil are so lively that they take care of things."

Biodynamic farming is relatively unknown in the United States, but in Europe, where Henning grew up, it is the primary form of organic farming. Henning's years as a humanities professor at the University of Washington are evidenced in his cogent summary of the history of Biodynamics. "In some ways this is old news," he explains. "Farmers have known this for centuries; they have just forgotten about it since Justus Liebig came along and invented chemical farming. In 1860, Liebig found that all we needed to do was put N, P, and K (nitrogen, potassium, and phosphorus) on the ground and we'd get these enormous crops."

"What Liebig hadn't thought about is that these chemicals, because they are highly soluble, kill the microorganisms and create toxic conditions that are inimical to earthworms." Henning continues, "So by the 1920s, the soils in central Europe were showing stress: falling fertility rates, falling productivity, increase in disease. Rudolf Steiner, who was a natural scientist and also a philosopher, was asked to help. Basically what Steiner did was systematize natural knowledge about building the micro-organic life

FARMING AND RANCHING

in the soil through these methods. He developed the system that is now known as Biodynamics."

Henning and his family have been growing their own food and practicing Biodynamics since he bought ten acres on Lopez Island in 1969. Henning describes their motivation succinctly: loving good food. There was also another factor. He explains, "As an undergraduate trying to work my way through school, I worked in a meat factory for a year. I learned everything I did not want to know about how animals were treated and the low quality of the product that came out of there. So I decided I either stop eating meat or I do it right. When I got my PhD, I immediately started looking for land so I could grow my own food."

Henning and his wife, Elizabeth, now own 15 acres and lease another 35. Striding purposefully through the fields, he describes what is growing this year on S & S Homestead Farm: a dozen sheep, two dozen beef cattle, a milk cow, three pigs, forty chickens. He reports, "We have about every fruit you can think of that will grow in the Northwest: apples, pears, cherries, peaches, four kinds of plums, raspberries, strawberries, gooseberries, red currants, black currants, blueberries—partly because we like good food and partly because it makes for biodiversity." About half of the food raised is consumed on the farm and the other half is sold, which is enough to make the farm economically viable.

Henning and his family use what their land provides. Building materials come from the farm whenever possible. Henning built the

farm's first structure 30 years ago out of logs he harvested from their land: a small windowless shack with a plastic roof to let in light. That's where he and his family spent their first years on the land, though the shack seems better suited to its current use as a very serviceable storage shed. The construction of on-farm living quarters has evolved considerably, but the materials are still close at hand. A pleasing straw bale bunkhouse for farm interns was built in 2001 from straw, clay, sand, and timbers harvested from the land. A pretty little Norwegian wood stove rests on white and blue tiles Henning's 90-year-old mother made and painted before she died.

Electricity is one of the few off-farm purchases, but here too, the farm is moving toward self-sufficiency. A cistern captures 2000 gallons of rainwater from the barn roof; the overflow is stored in a pond, and from there the fields, gardens, and orchard are irrigated using a solar power pump. Animal waste is carefully composted and spread back on the ground. "It's a very complex system that takes a lot of management, but we have been doing it for 35 years now, so it's a well-oiled machine," says Henning.

Henning is committed to passing on what he has learned living on the land. He serves as an adjunct professor for Washington State University's sustainable agriculture program. Most of the instruction and research happens on his farm, though he also lectures around the state. Henning and Elizabeth also teach a course in ecological food production to local high school students; students help grow, harvest, and prepare greens that are served in the school cafeteria. Students are also involved in a research project evaluating

21 varieties of heirloom beans to determine which grow best in the Lopez soils and climate. "If we can find the strains that are best suited," he reports, "we can develop niche markets. You get 89 cents a pound for beans on the commodity market, but you can get $2.89 a pound at a farmers' market."

About 40 percent of the family's income is from education; the rest is from custom meats, vegetables, and dairy. The farm has 10 CSA subscribers for vegetables right now, and plans are in the works to bump that up to 50 with the help of a farm manager/education coordinator. Henning reflects "People always ask me, 'can you make a living doing this?' The way we think about it is this: not only what are our receipts, but also, what did we produce internally, what don't we have to buy? The fact that our transportation costs are one-seventh of the national average is a significant savings. The fact that we have not been sick in 30 years is an important savings. The fact that we don't have any chemical or veterinary expenses is

an important savings." Explains Henning, "Our cash income is relatively low, but when you actually internalize the benefits, it is easily equivalent to what the Department of Labor considers an average family income."

There is much more to Henning's food system than the economics. He explains, "A lot of people are so used to commercial food they can't taste anything that is subtle or complex. They ask, 'Is it sweet, is it greasy, is it salty?' And that is all they taste. The vegetables and the fruit here have a whole symphony of flavors. I think that is a result of the way we manage the farm and the soils. If you buy an apple in the store it is either sweet or not, but it doesn't have this range of taste." Henning concludes, "The diversity of this place is reflected in the flavor, the aroma, and the texture of the food we produce here—in the apple, the carrot, the potato, and the meat and milk."

BUSINESS

Over the past few decades, rural businesses have had to contend with greater consolidation and globalization in the marketplace. To survive this change, small businesses are finding ways to diversify and specialize. Some have moved toward vertical integration, bringing packaging, processing, research, and marketing in-house. Others use the internet and creative marketing strategies to sell directly to their customers. Many have embraced social responsibility, investing in their communities and providing better conditions for workers and their families.

Regardless of the strategy, it is clear that today's successful rural enterprises must do more than rely on the status quo or the bustle of Main Street. Successful small businesses are finding they must either increase their own sophistication to compete in the global marketplace, or focus on niche markets and value-added products that attract discerning consumers.

These stories are a testament to the ingenuity of rural-based entrepreneurs who are holding their own against the growing tide of globalization.

INABA PRODUCE FARMS

Inaba Produce Farms is a large, family-run operation in the arid Yakima Valley that grows, packs, and ships a diverse variety of vegetables throughout the region. The Inabas have been experimenting with organic methods in an effort to improve the farm's diversity and to reduce their dependence on chemical inputs. It is not an easy transition to make, but Inaba family farmers have survived challenges here before.

"We farm about 1200 acres, and it's a family operation," says third generation farmer, Lon Inaba. "I have two brothers: Wayne is our

salesman, he's the money guy; Norm is our computer guy, he has a degree in economics and computer sciences, and does our payroll and taxes. My mom is 75-years-old and she is our office manager: she pays the bills, collects the money, does the shipping papers, and keeps us all in line." Lon himself is the engineer, innovating and developing new things. He builds the greenhouses, drip irrigation systems, and composting operations.

Inaba Farms is made up of wide, flat fields, broken only by power lines and the occasional road. The surrounding hills are faintly visible in the distance. It is not hard to imagine the sagebrush country this once was. It takes several minutes for Lon to drive the entire length of a compost windrow in one of the fields. "We have about five miles of compost windrow," Lon remarks. "We started composting as a way to build up our ground. We've used manure and cover cropping for probably the last 25 years; my grandfather used those practices almost 100 years ago." Utilizing their waste products in the compost pile helps the Inabas reduce their chemical use.

Lon speaks familiarly about the relatives who worked this land over the last century. He says, "My grandfather came from Japan in 1907 to the Yakima Valley because this was one of the last areas still in sagebrush and they were asking people to immigrate here to help break and farm the ground." His grandfather broke 120 acres out of sagebrush, some of the ground they farm today. Lon recounts, "My grandfather was doing pretty well, but then the '20s came around and they passed the Alien Land Law that said that 'undesirable aliens' couldn't own or rent land." Lon's grandfather was reduced to

sharecropping, and could no longer afford to rotate high value crops with soil-enriching hay. He had to move every couple years when the soil was worn out. This forced him to look for a high margin crop that he could justify with fewer acres, which is how the family got into the vegetable business.

Lon recounts, "They had moved to eight or ten different places before my dad's cousin was old enough to sign for land. And by that time World War II came around and the Japanese bombed Pearl Harbor and there was a lot of animosity toward the Japanese people." This time, world events meant that Lon's grandfather, father and mother had to move to an internment camp for several years. "So we've seen adversity," he says, "and we can relate to our workers, most of whom are Hispanic migrants. Just like my grandparents, they gave up everything they had to come to this country to try and make a living."

Lon left an engineering job to come back to the farm 20 years ago, when his dad decided he wanted to get into packing and selling his produce on his own. Lon recalls, "I took a six-month leave of absence to help build a cooler and a warehouse and I have been here ever since. I did more actual engineering during my first three weeks home on the farm than I had done in the previous three years!"

In 1982, the farm consisted of three crops: concord grapes, bell peppers, and sweet corn. Today the Inabas produce nearly 20 different crops. "We try to do everything from start to finish and have something the consumer can put on their table," explains Lon. "In our business the margins are pretty slim. If you take what the

consumer pays and divide by five, that's about what the grower actually gets. People don't realize that. We have to be vertically integrated to make this work financially." The Inabas grow their vegetable starts in greenhouses and have their own packaging plant. Merchants can come to the farm to pick up four or five things at a time and that helps them attract customers.

The Inabas farm both organically and conventionally, about 10 to 15 percent of their crops are in organic production. Lon says, "We do a little bit of almost everything organic. We're pretty diverse to begin with and the organic thing makes us more diverse. We sell to major retailers in the Pacific Northwest and to a few throughout the western United States and Canada. The organic deal gives us more specialty items to sell."

The harvest season begins in April with asparagus; then they move into cabbage, peas, green beans, cucumbers, yellow squash, and zucchini; later comes sweet corn, bell peppers, watermelons, onions, tomatoes, and eggplants. Lon explains, "Our workforce is about 150 to 200 people and probably two-thirds of the same people come back every year. We get good people, and we try and take care of them."

we started looking at using less harsh chemicals in the remainder of our crops so we wouldn't kill the bees." Lon also is concerned about the health of his workers in regards to these chemicals. He adds, "My dad taught us: you treat people the way you want to be treated. If we don't want to spray or work around chemicals, we don't want our workers to either."

After the Inabas tried numerous organically-certified chemicals, most to no avail, they decided to try different management practices and beneficial insects. Lon explains, "In our organic fields, we're trying to be diverse in our cropping structure and diverse in our management practices. I try to release beneficial insects early so we have second and third generations of these insects throughout the growing season. I release lacewings, parasitic wasps, midges, and ladybugs. Hopefully they won't fly away. Hopefully when we have a problem, they'll be there to eat the undesirable guys." Lon creates habitat for beneficial insects by walking through his fields and tossing handfuls of clover, alfalfa, yarrow, corn, wheat, and dill seed, providing diversity in the pollen and nectar sources for the insects he releases in his fields, as well as attracting other beneficial insects.

Inaba Farms has built quality housing for migrant workers and has even developed housing sites among the fields that they lease for a nominal fee to some of their most valued employees.

Lon also works to make sure Inaba Farms is a good place for insect families. "Squashes, cucumbers, and watermelon are crops that need bees for pollination," he explains. "As we started adding those crops,

Lon seems to relish the challenges of farming with fewer chemicals, but it clearly is not easy. He says, "A lot of the stuff we're doing for organics we do on faith because there is no instant result. You have to hope the beneficial bugs will stick around and that the decisions you make are going to be good—there is no instant verification of your results." He continues, "I do a little bit of a lot of different things—the diversity approach—and hope that some of those things are working."

LOCATI FARMS & WALLA WALLA SWEET ONIONS

Walla Walla Sweet Onions have been grown for over a century on small farms nestled between the Columbia and Snake Rivers and the Blue Mountains. These onions have half the sulfur and more water than other yellow onions, and they don't make you tear when you cut into them. Sweet onion seeds were first brought to this area from the island of Corsica. Italian immigrant farmers, impressed by their winter hardiness, began cultivating them and selecting onions for seed based on sweetness, large size, and round shape.

Michael Locati's grandfather started growing sweet onions here when he emigrated from Italy in 1905. Michael grew up on an onion farm, and he describes farming as a kind of disease he tried to avoid. "Once you get into high school, you think you are smarter than everyone else, and you think there is a better life out there, so you go out and you explore a little bit," he laughs. Michael worked a few jobs in the agriculture industry but soon realized he needed to be back on a farm. He comments, "But you don't just go out with your checkbook and say 'okay, I want to farm, here's $500,000; lets do it.'" Michael worked for 12 years as an electrician, and worked his way up to being a contractor, but still dreamed of farming.

In 1979, with no equipment, almost no land, and no money, Michael started a little onion seed business on the side. He explains, "I just had an idea that I would start farming. I started with one quarter of an acre of Walla Walla Sweet Onion seed. It turned out growers here in the valley were having a hard time producing their own seed, and they started buying the seed from

me." The business grew from there. By 1987, Michael was a full-time farmer and a full-time electrical contractor; he soon decided to sell the contracting business and just focus on farming. He reflects, "My heart was in the farming—so I don't have any regrets."

Michael's farming venture grew and he started securing the option to purchase the land he was leasing. Fortunately, some landowners were willing to carry low interest contracts for deed, which made it financially feasible for him to buy the land. "It's a nice way to hand these farms down because in some cases the children aren't interested in farming and these landowners want to

keep the farm intact if they can," he explains. Michael's farm is now close to 500 acres.

Michael has developed a three-year crop rotation to cultivate soil fertility and minimize weed and pest problems. He has also diversified crops to avoid disease problems, chiefly the incurable white rot that is common in the valley. Michael explains, "We end up with ground that we can't use for fall onions anymore. So we'll put in crops like alfalfa and hay because they are in for five years and we are not tilling the soil. We want to seal that soil down and not contaminate more fields." He plants asparagus for the same reason—because it's a perennial crop and once it's in you don't need to till the soil. The asparagus business also provides additional work for Michael's employees: "It keeps our crew busy. We start up in April and keep them until August, and without asparagus we don't start up until mid-June." Locati Farms has five seasonal full-time employees, and twenty-five seasonal part-time employees.

Dust and mud are two of Michael's least favorite parts of farming. The other thing he dislikes is pesticides. He says ardently, "I don't like dealing with them; I don't like my employees dealing with them." According to Michael, as various pesticides have been taken off of the market over the last two decades, other farmers have said, "How am I going to survive?" He says, "Good riddance! This has opened the door for other things. I have a lot more tools now that I can use that are safer for me, safer for the environment, and safer for my employees." He continues, "I am not an organic farmer and I am not a conventional farmer. If I have to bring out the big guns, I'll bring them out, but I try and do everything I can to prevent that situation from happening." Michael has been experimenting with biological controls, gentler pesticides, and minimum till. As a not-quite organic and not-quite conventional farmer, Michael is certified by Food Alliance and sells his onions and other produce under that label.

In the early 1990s, the onion growers in the Walla Walla Valley were in danger of losing their market share to large farms outside the area. These operations could sell their onions at a cheaper price because of their scale of production and mechanical harvesting. "People would complain, 'These don't taste the same,'" recounts Michael. "Sometimes they would have a decent crop and then they'd put in three more circles. Well, that equals what all of our little farms do here in the Walla Walla Valley."

Michael felt that if they didn't do something, the small Walla Walla onion farmers would be history. He comments, "It would be, as agriculture often is, that bigger, stronger growers take over that particular market and drive it. It would still have our name—it may get trashed, but it would still be out there—but we wouldn't have a position in it anymore." He continues, "We have small fields here and our harvests are done by hand. Our costs are higher; there is no way we can compete with large-scale farming."

Michael came up with an idea. He helped found the Walla Walla Sweet Onion Growers Association and set to work defending the Walla Walla Valley's onion legacy. He explains, "The Walla Walla Sweet Onion name wasn't protected, we didn't have a trademark or a geographic area. I felt that we needed to protect our industry." For two years Michael strategized with seven other growers, crafting a federal marketing order that was submitted to the USDA for approval. In 1995, the order went into effect, protecting the geographic region and the name. Now, only growers within a 15-mile radius of where sweet onions have traditionally been grown and marketed can sell their sweet onions under the Walla Walla name.

Cooperating with other small farmers has been an effective way to stay competitive. When the marketing order came into effect, Michael started building his own packing facility and by 2001 began packing all of his own onions as well as those of a couple of other growers. Then he merged with another company in Walla

Walla and built a new packing facility with eight other partners. Michael and his partners now own one of the biggest packing facilities in the Walla Walla Valley and produce about 60 to 70 percent of the Walla Walla Sweet Onions.

By working together, the relatively small farms in the Walla Walla Valley are able to achieve some of the economies of scale available to larger producers. Michael explains, "When I was getting into the industry, I was a grower, packer, shipper, sales person, floor sweeper, everything. Now my full responsibility is growing. I am not running a packing facility; I am not on the phone trying to sell my product anymore. It's being handled by a partner that is in that line of work and I trust that they'll do their job and they trust that I'll do mine. We have a good partnership."

As for the consumer, now when you buy a Walla Walla Sweet Onion anywhere in the country, you can be assured that you are getting an onion that carries with it the flavors and traditions of the Walla Walla Valley.

MIKE AND JEAN'S BERRY FARM

Mike Youngquist's Swedish great grandparents homesteaded in the fertile Skagit Valley in 1889, clearing six-foot diameter cedar trees so that they could till the land and grow oats and hay to fuel Seattle's horse-drawn buggies. Today, Mike and his wife Jean face challenges their ancestors couldn't have imagined; they have adapted to the increasingly competitive market for small farmers by becoming highly strategic business and political operators. This means they spend less time in the field and more time processing and marketing the strawberries, raspberries, cucumbers, peas, and cauliflower grown on the family land—and making trips to Olympia and Washington DC.

Mike grew up in the dairy business, but developed allergies to grain dust and grass pollen and made the switch to growing berries. Jean also grew up nearby, and, like Mike, spent her summers working in the fields. Mike and Jean have seen dramatic changes in agriculture in the Skagit Valley in their lifetime, and they talk about those changes the way other people talk about their families. Says Mike, "The glory times of this valley were during the '60s. There were eight or nine processing plants. It was one of the major frozen green pea producing areas in the country. But now that the processors

have left, most of the growers have gotten into value added agriculture to stay in business."

For Mike and Jean, this meant investing in processing equipment for their berries and other crops. It has also meant that they have had to become vertically integrated, which means shifting from just growing and delivering a product to processing it, packaging it, planning for a continuous supply, and being responsible for all the marketing and account management. "It's a whole different ball game," says Jean, "and the shift takes a large investment in overhead, including cooling facilities and processing equipment."

Strategic adaptation has allowed Mike and Jean to remain economically viable and carry on their family's farming tradition. Says Mike, "We started out as a production company; then we went into processing, and it took us ten years to learn how to do that. Now we are into marketing. Each job is specialized, and you go through severe growing pains."

Consolidation in the food industry has meant that grocery store chains save time by buying from fewer, larger farms. Mike explains, "There is a monopoly on food distribution channels and in the food service industries. Four or five industries serve the US, and, when it comes to grocery store chains, again four or five control the majority of the market." For Mike and Jean, it is not the price or delivery that is the obstacle to selling to these chains, it is getting the chains to purchase in smaller quantities. The Youngquists have worked hard to get to know buyers, who often are located halfway across the country. Although it's easier for those buyers to connect with large firms supplying a little bit of everything twelve months of the year, Mike and Jean have successfully convinced buyers to purchase on a regional basis.

Fighting the consolidation and globalization of the food industry is a personal mission for Mike and Jean. Jean has been lobbying for national legislation to require tax-supported institutions, like schools and hospitals, to serve American-grown food products and thereby strengthen the viability of a safe, domestic food source. And a consumer-driven shift also seems to be working in their favor. Mike explains, "Some firms are getting smarter and are buying locally because consumers are demanding that their fresh produce be grown closer to home. The trend was to buy nationally, but people in this country are beginning to demand food that is local, clean, and safe."

Mike and Jean have developed innovative channels for selling their produce directly. They do fundraisers with Rotaries, Kiwanis, schools, and even a museum. Groups take orders from members within the community and then buy fresh processed berries from Mike and Jean. With a mark-up, these groups are usually able to

earn ten dollars per pail and still sell under the retail price. With every pail sold, somebody—kids, seniors, or even parks—benefit.

The Youngquists are members of Food Alliance, a non-profit labeling and marketing program that certifies sustainable agricultural operations. Mike explains, "We were looking for a way to niche

market and the advantage of going with the Food Alliance is they're actually making agreements with store chains. It gives us salespeople on the outside and allows us, as a smaller operation, to join with many other products." He adds laughing, "We like the Food Alliance because it's organic without the religion."

In the 1980s, changes in child labor laws and the advent of on-farm processing signaled the end of busloads of local kids picking

BUSINESS

berries and frolicking on the farm during their summer vacation. Mike and Jean are the first generation of Youngquists to rely on seasonal migrant workers. Many of the same migrant families come back to Mike and Jean's year after year, and some families have been returning, and bringing their relatives, for nearly 30 years. Seven years ago, the Youngquists created an award-winning daycare and school for the migrant families working in their fields.

"When we started The Berry Good School, there was no place for the kids to go during the day," says Mike. "We couldn't allow the kids to be in the fields if they were under 12. So kids were left alone, unsupervised, and that wasn't good for anyone. We started this daycare to provide a convenient alternative, and since it's associated with the workplace, people have more respect for it and feel comfortable leaving their kids there."

Mike and Jean are family farmers beating the odds by evolving with the marketplace, and lobbying consumers and policymakers for support of domestic food sources. "Consumers in this state have to buy locally even if it costs a little bit more," Mike says. "By paying a little more for their food they are also helping to maintain the open space and habitat that farmland provides."

The school serves older as well as young children, and features a state-of-the-art computer learning center. With this jumpstart, Mike and Jean hope that these kids will have more career options than their parents. "We feel that these people are part of our society," says Mike. "They are going to be productive American citizens and taxpayers, and the quicker we can treat them as equals and give them opportunities, the better off we all are."

He continues, "If you follow the economic structure of always buying from the lowest priced source, agriculture will leave the United States because our land is too expensive and because of our recreation demands, our labor costs, and our environmental regulations." Mike concludes, "We have the cleanest water and the cleanest food in the world, but it is expensive. Consumers in this state are going to be the ones—voting with their pocketbooks—to decide whether we are going to survive or not."

PARADISE FIBERS

Kate Painter holds a PhD in agricultural economics, but her farm-based business is driven more by her passion for fiber arts than her knowledge of economic theory. Kate is teaching her daughters to knit, carrying on a family tradition from her Scottish heritage.

Kate became enamored with life in the country when she was a teenager and her father used a layoff as an opportunity to follow his dream of becoming a farmer. "I thought it was a great adventure," she remembers. "We had cattle, horses, and a hay and grain operation in eastern Washington. My biggest purchase as a 12-year old was a milk cow!" As an adult, Kate and her husband, Charles Knaack, purchased a small farm in Whitman County so Kate could have a sheep operation.

Both Kate and Charles worked at Washington State University, until they became parents. Then Kate's career aspirations shifted. She explains, "When I had my two little girls, I thought if I could make a few hundred dollars a month, I could stay home with my kids while they were little." She continues, "I had all these sheep fleeces and thought I would start a little hobby business—a mail order company with a catalog." That was in the mid-1990s and in her first year she had about $6000 in sales. Last year Kate had over $150,000 in revenue, mostly made up of small orders.

While Kate still sells the fleeces from her flock of sheep through her website, they now account for only a fraction of total sales. Paradise Fibers offers knitting and spinning supplies from all over the world, along with yarn, needles and fibers for spinning. Interested shoppers can also buy a wooden spinning wheel straight out of Rumpelstiltskin, or intriguing-sounding tools and accessories like whorls, swifts, and niddy-noddies.

Kate attributes part of her success to the computer knowledge she gained working as an economist. "When websites came along it was natural for me to put my catalog online," she explains. "I started putting pictures of my fleeces on the web and realized that people love the feeling that they are looking at a farm, with pictures of the animals." While it is time consuming, Kate does send samples of the fleeces through the mail so customers can touch them and see how they spin. However, she is amazed at how many people are willing to make purchases just from looking at a scanned picture.

"You develop a reputation being on-line with mail order. It's important that people know you are trustworthy," she explains.

A web-based business works well for someone who needs the flexibility to look after animals. Kate briefly tried having a store, but found it wasn't workable during lambing season when she needed to be home. Besides, all of her growth was coming from Internet sales. "I can't believe how many new customers I get, six to ten a day, all from the Internet," she remarks. "It's a very specialized market. My biggest expense is advertising in magazines to let people know I am here. The Internet is amazing. The problem is you really have to be a jack-of-all-trades to do the marketing and the business end; I am always juggling."

Kate's basement is lined with tidy shelves of color-sorted fleeces and yarns. She has found it is not economically feasible to sell yarn spun from her own sheep's wool, but she has found other ways to make creative products. Kate buys leftover fiber from large yarn production companies and mixes them with other fibers—mohair, yak down, or silk—to create unusual, luxurious blends for hand spinners. These unique "microblends" are her most profitable offering. Kate imports interesting fibers from across the globe for hand spinners, like leftover silk from sari production in India. "People have been going crazy over it," she explains, "and there is enough margin I am able to wholesale it through stores."

Kate occasionally hires someone to help with shipping, but most of the work she does herself. Her best friend developed and produces her hand-dyed wool line, called Rhapsodies. Eight ounce rounds of coiled fleece are dyed a harmonious assortment of colors with evocative names like Monet's Garden, Harvest Sunset, and Winter Solstice. Kate leaves most of the farm work to her husband, who fortunately enjoys feeding animals, keeping up the fences, training border collies, and maintaining pastures.

While her sales are impressive, Paradise Fibers is not making Kate wealthy. "For years and years—and to some extent to this day—I feel like I subsidize my business by not making a living wage," she says. "Sometimes I get discouraged, but as I tell my kids, there is a lot more to life than having lots of money." For Kate, the time that she has with her family, and the experiences they share living on a farm and raising animals and their own food, is well worth it. She comments, "I am pretty independent and I would rather buy my clothes at the Goodwill and drive an old car for the tradeoff of doing whatever I please. In this country it is pretty easy to live frugally if you are not picky, and I would rather have free time. I think it's good for kids to grow up with less and realize that resources are limited."

Paradise Fibers dedicates two percent of its profits to promoting fiber arts, particularly among Native artisans. Kate has sent yarn, fiber, and household goods to Navajo weavers. She recently sent a spinning wheel to India and collected donations of over 200 pairs of knitting needles to send to Ecuadorian knitters. Kate also supports other artisans from around the world, buying products like handcrafted knitting needles from Nepal that are carved from rosewood and bone in the shape of frogs, owls, and elephants.

Kate hopes that the resurgence of interest in knitting will continue. Certainly, Paradise Fibers' website offers much encouragement to those who are considering exploring the textile arts. In addition to supplies, there are books and patterns for sale, knitting and spinning videos to rent, helpful hints, an online newsletter, a calendar of related events, and even a quarterly trivia contest. "It's my hobby, and I just love anything to do with the traditions and old textiles," says Kate with a smile. Walking in the pasture with her fuzzy flock of brown and white sheep close at hand, Kate seems content with the life she has woven out of her passions for farming and fiber arts.

SAKUMA BROTHERS

While a lot of farmers talk about the need to be vertically integrated, the Sakuma brothers have fully embraced this concept. They still sell fresh strawberries, raspberries, and blueberries at a market stand during the summer months, but that represents only a tiny fraction of their business today. Over three generations, the Sakuma family has built a successful fruit business that operates in two states and includes a farm, a research and development laboratory, a wholesale and retail nursery, and a commercial fruit processing and packaging plant.

Steve Sakuma is the current president of Sakuma Brothers. His grandparents immigrated to Bainbridge Island from Japan in the early 1900s and set to work, farming and raising ten children. Steve's father, the first of his grandparents' children, helped the family take products to Seattle, a ferry ride away. The return was little more than a hand to mouth existence, but it was enough to feed the family of 12. In 1935, a window of opportunity opened. The cannery in Seattle to which they delivered their strawberries sponsored the Sakuma family's move to Skagit County. It was an opportunity to own land, to farm, and to be more independent inside what was emerging as a very corporate economic system.

In 1941, with part of the family on Bainbridge and part in Skagit County, the Japanese struck Pearl Harbor. The Sakumas, like all Japanese families in the region, were driven out of their communities, first from Bainbridge and then the Skagit Valley. The latter evacuation permitted them enough time to get things in order, including appointing a power of attorney and finding a caretaker for the land.

At the end of the war, the Sakuma family returned to their farms, but found they had lost their land on Bainbridge Island. They were able to retain their land in the Skagit Valley, however, and started farming again in 1945. With a new sense of purpose, they took to the fields and began selling strawberry fruits and plants. The family decided to start a nursery, but a certified nursery requires that two fields lie fallow for every one in production, and that much land was hard to come by in the Skagit Valley. To expand, Steve's father and uncles moved the strawberry nursery operation to northern California in the 1960s.

In Washington, the Sakuma family expanded into raspberries and blueberries. Steve recalls, "In the late '80s and early '90s, all the processing facilities for small fruit in Skagit County just went away: they couldn't make it. The paradigm of agriculture was changing at that point: if you weren't a grower and a processor, the future wasn't there." In 1990, the Sakuma Brothers built a processing building for a company from Oregon; in 1997 when the company left, the family bought the equipment. With the addition of the processing operation, the vertical integration was complete, but they didn't stop there.

The addition of a small fruit research lab means that Sakuma Brothers are on the cutting edge of propagation technology, relentlessly pursuing disease- and virus-resistant alternatives that provide high yield and superior fruit. The whole package is unique in the small fruit industry, and it means that Sakuma Brothers is a player in national and even international markets. "It's still the structure of a family business, just multiplied over time," explains Steve.

He continues, "The process begins with the potential customer, taking them into the lab and showing them where the nursery operation starts, then out to the fields, and eventually back to the processing plant. It gives the potential customer insight into a few things: We are here to stay, we are invested in this operation, and we have the potential to grow."

In Washington, Sakuma Brothers currently harvest 160 acres of strawberries, 300 acres of conventional and organic raspberries, 300 acres of conventional and organic blueberries, 57 acres of apples, and a fledgling conventional and organic blackberry business. "It's apparent to us that if you really want to work the entire market, you've got to be in both conventional and organic," explains Steve. "We have found that organic really supports conventional and conventional supports organic. We wouldn't have been able to take on organic if we weren't reasonably good at conventional farming because you've got to have a basis for it. And now we have found that what we have learned through organic farming is strengthening our conventional farming."

Steve is proud of the success of this business model, but not perhaps for the reasons that one might expect. "What is really important to us at a corporate level is our family legacy," he says. "The business is very important, but the reason the family name comes before the business is the family is more important than our business. We believe if we take care of the family, the business will follow. It's worked through two generations."

In 2004 the company brought on their first fourth generation member, Steve's son. A cousin from the third generation also joined the ranks as the first woman at the corporate director level in the company. Steve explains, "Our family was pretty male dominated in the second generation, but as we looked down into the fourth generation it became very apparent to us that if we didn't recruit our daughters and nieces, we wouldn't be able to maintain a family business."

Recruiting and training family members to carry on the business has become more complicated than it used to be. "When we were kids," Steve recalls, "if you were three years old and you were old enough to stand up, you would go out in the field. We grew up out in the field." Now child labor laws prevent children under the age of 12 from working in the fields. However, farm families are able to have their own children work on the farm. "It gives us an opportunity to establish what we believe is the right work ethic," explains Steve.

The Sakuma's values include choosing to take an active leadership role in sustaining agriculture. Steve has served as the president of Skagitonians to Preserve Farmland, sits on the Agriculture Association Board, and is a drainage commissioner in his district. "We believe that if agriculture is going to be sustained in this valley, we have got to participate," he says. "We have to be part of that education process. We have to be sure the right decisions are being made by county, state, and federal officials."

Most of the fourth generation Sakumas are now in college, and time will tell if they choose the path that has been forged for them. The family's values of hard work and loyalty are firmly ingrained in the older generations. "Our business is based on the assumption that we are going to be here well into the future," states Steve. "Because of our parents we have what we have. Now it is up to us to take what was given to us and increase it to the point where we can pass it on to the next generation."

SHEPHERD'S GRAIN

Today the global commodity system ensures that almost all Northwest-grown wheat is sold and milled abroad. The typical wheat farmer in the Northwest plows the fields every fall, sows a soft white wheat seed in the spring, hopes for a good commodity price when it's time to harvest in the fall, and then buys imported flour at their local supermarket. Shepherd's Grain is finding local markets for locally-grown wheat, turning the conventional paradigm on its head.

They have also eliminated the plow. Shepherd's Grain producers are proponents of no-till or direct seed agriculture; they believe in preserving the integrity of the topsoil. Shepherd's Grain fields are never left bare, so dust storms and run-off are greatly reduced. When it is time to plant, a machine called a "drill" punches new seeds into the remains of last year's crop. A growing field of direct-seeded wheat has fuzzy strips of tan stubble interspersed with new green sprouts.

Fred Fleming's family has been conventionally farming for four generations. He admits that it was hard to make the transition to no-till farming, but he could see benefits from the start. Fred recalls, "The very first year that I was direct seeding, I was standing out in the field when we had a tremendous rain come through. It must have dumped a good inch in 15 to 20 minutes. My neighbor—who was a conservation farmer of the year—had just planted, and his soil was washing into the aquifer, boiling into the culverts, pushing up against the road bank. My farm was just starting to show a little trickle, it was sponging it up, holding all the water."

Fred and his business partner, Karl Kupers, live and farm in eastern Washington, just below where the Columbia River veers north into Canada. Karl remembers when he and Fred decided to leverage their enthusiasm for no-till farming into a business venture. He recalls, "Fred and I got together to talk about how we could utilize the Food Alliance and their concept of third party verification for a direct-seed system." The Portland-based Food Alliance certifies sustainable farmers using an array of human and natural resource criteria. Getting Food Alliance certification was the first step, followed closely by product development and a search for viable markets.

Fred and Karl learned quickly that accessing markets is not an easy process. Says Karl, "The first two years it was just Fred and me with a vision. We went to five specialty markets in Portland and we saw those cylinders of bulk product—wheat, flax, flour, sunflower, mustard—all sorts of products that we could raise." Karl tried to talk his way in, with a handful of no-till sunflower seeds in his pocket, but the staff politely explained that he could come back on the forth Tuesday of the month and wait in line with the rest of the farmers.

So Fred and Karl drove 300 miles back home, regrouped and tried another approach. They decided to focus on what they do best:

grow wheat. They raised it, milled it, and then brought it to artisan bakers to try. They focused on Portland, since Food Alliance is based there and had many connections to share. They tested different wheat varieties, and two rose to the top. Karl recalls, "A baker tried blending them and together they make a really good, flavorful and functional flour."

Identifying that flour blend has been no small part of Shepherd's Grain's success. Laughs Karl, "Fred and I aren't smart enough to know exactly why we ended up with this great product, but we did. As Fred says, 'Perseverance and ignorance will prevail!' We were told you can't grow a good dark red spring wheat in the

Pacific Northwest; we were told it won't make good bread flour; we were told it will never compete with hard red winter wheat for a flour. We have debunked all three."

Karl and Fred are also breaking new ground in how they price Shepherd's Grain products. Wheat is typically a commodity product, meaning the price is set by the Chicago Board of Trade. Karl explains, "Commodity pricing is necessary in the global arena, but from a domestic standpoint, we don't need it." He remembers the meeting where they decided to use a different approach, citing studies that show that sustainable, local, safe, and traceable food is desired by consumers. He says, "We had a baker, we had a miller, and we had us, and we talked about this whole project. We all agreed that it had merit and that we should go for it." Karl continues, "But of course to be truly sustainable, it has to be economically sustainable."

"The first thing we were asked," Karl remembers, "was how we were going to price the raw product?" Fred, a long-standing commodity broker whose office is all based upon the Chicago Board of Trade, immediately said, "We're going to take the Kansas City hard red winter futures in November and we're going to add a 50 cent premium." Karl knew he would be the one out there marketing the wheat and he wasn't sure how he was going to substantiate such a premium. So he asked the baker, "How do you price your bread?" The answer: "Cost of production plus a reasonable rate of return." Then he asked the miller the same question and he replied: "Cost of production plus a reasonable rate of return." Karl decided that answer should apply to their product too.

The upside for Karl, as Shepherd's Grain's primary marketer, is the transparency. He explains, "I can look a customer in the face anytime, anywhere and tell them that we have an honest price. It's not asking for Cadillacs and trips to Hawaii. If buyers argue about

the price, then they don't want the farmer, the miller, or any of us out here to be in business next year to do this again, bringing that same quality product—not only next year, but the next generation."

Fred and Karl are very proud of what the Shepherd's Grain label represents. Fred explains, "If a consumer wants to be an activist, they are doing more for the environment and society, in the long run, with this one purchase. With the no-till or direct-seed method, the soil stays here on the farm, it doesn't get into the streams where it can cause harm." He adds, "This label represents locally grown, family farmers. When you become this kind of food activist, you also become a disciple to save family farms."

Fred describes himself as a conservative hippy and is quick to point out that this venture has not been for the faint of heart. It is driven by passion, and it has to be. "You have to have the willingness to come to the new land and burn your ship so there's no way of going back home. That's really what you have to have to make something like this work," he laughs.

There is also a push on the demand side. Karl explains, "Part of this whole vision was that we could see that other countries that used to import our products—specifically from the Pacific Northwest—are now net exporters." Fred adds, "All the farmers out here hope this works and have been very supportive. They have seen this cloud of change that is on the horizon, just steaming down the valley at us. So far we haven't been affected, but it's coming. Maybe we will only carry the ball so far down the line, and there will be that next group that will take the ball and score the touchdown on this, but we're hoping we'll score the touchdown. We're trying to create a new paradigm."

THUNDERING HOOVES FAMILY FARM

Joel Huesby doesn't look much like a radical. He's a native eastern Washington rancher sporting the standard jeans and cowboy hat. When he starts talking, you quickly notice that he seems unusually animated and voluble for a rancher. As phrases like "sunlight farming" and "active microbial life" start cropping up thick and fast—interspersed with raucous laughter—you realize that this is a different breed of rancher altogether.

Thundering Hooves sells pasture-raised beef, chicken, and turkey directly to consumers and restaurants in the Seattle and Walla Walla areas. Joel has a grill in his butcher shop, and if he is uncertain whether a particular cut is worthy of steak, he'll cook it up right then and there. "Because we are small, efficient, and flexible, we can move animals into various products depending on how they taste," he explains. Joel's animals are raised without hormones or antibiotics, and roam on green pastures that have not seen chemicals or pesticides in years. But it wasn't always this way.

The Huesby family has worked land in and around the Walla Walla Valley since 1883. Joel says, "I have been a farmer essentially all my life. I grew up around cattle and crops, with chemical farming and tractors—the whole nine yards." Joel grew his first field of wheat when he was in high school, and made enough money to buy himself a pickup truck. But by the time he was starting his own family, the profit margins had dropped dramatically. Joel was farming the way his uncle and his grandfather had before him, but it wasn't paying the bills.

Twelve years ago, Joel suddenly veered off the conventional farming path, without knowing where else to go. "My epiphany moment came when I was burning a field of wheat stubble; I was going to plant alfalfa that fall. Watching the smoke go up, something clicked in my mind." Joel realized there was a fundamental flaw in his farming model: the chemical inputs he was putting on his fields were yielding diminishing returns. He began to do his own research and decided that soil fertility was better achieved by returning organic matter to the soil through grazing livestock and growing legumes. But it wasn't an easy transition to make after years of conventional farming practices.

"I talk about conventionally-farmed soil as a drug addict," he explains. "Without its next chemical hit it will fail to produce. Just like with an addict, it's progressive and it's terminal. And for the farmer, it's ugly. There is a period of withdrawal. Weeds come, the soil is lashing out. And there is only one way to do it, and that is just enduring the time it takes to go through it. You will find life on the other side, but it can take several years—I know!"

During his soil's rehabilitation period, Joel worked part time installing phone lines and computer networks with his father. He also tried his hand at horse farming, but after a spectacular accident that involved broken power lines, water mains, and skin and bones, Joel switched to smaller livestock. The big draft horses did leave their legacy in the farm's name, Thundering Hooves. Today the operation includes a slaughter and meat processing facility, and employs Joel, his wife Cynthia, his brother Bryan, and four other full-time employees, plus part-time help from Joel's sister, parents, and kids.

While the Huesby's are fully wired when it comes to marketing and communications, relying on computers and the Internet to correspond, their farming techniques are less high-tech. Joel says, "Here on the farm we have embraced all the things that were good and productive about the traditional family farm, including diversity of livestock and putting animals back to work for us." Joel uses older tractors and an immaculate egg incubator from the 1940s that wouldn't look out of place in a living room.

Joel employs "biological efficiencies" on his farm. One example is the way he feeds his cattle. Instead of cutting, drying, baling, and hauling alfalfa to his cows, he lets them do most of work. Joel mows a field, allows it to wilt a bit, and then lets the cattle feed themselves. He remarks, "This is a big no-no for most cattlemen because you can loose a lot of animals to bloat that way. But I have found—through a few tricks and the school of hard knocks—you can get incredible gains in productivity letting them harvest their forage themselves. God gave them four legs and a mouth, let them go harvest their own!" Joel uses electric fencing to intensively graze small portions of pasture for short periods of time. It allows him the added benefit of having the livestock spread their own manure.

One challenge that Joel has encountered is that today's livestock have been bred to fit the industrial model, which selects for average daily gains, high weaning weights, high yearling weights, and feedlot performance—all on a high-concentrate, high-intake diet. Explains Joel, "But I am more interested in pounds of steak sold per acre grazed, not per animal, so I am more of a natural resources manager. I am using cows to do the harvesting. The cattle in my grandfather's day were smaller and proportionately shorter, wider, and longer; they fattened on grass much more easily and more quickly. I have a long-term goal of bringing back the kind of genetics that works with grass."

Four years ago, Joel figured out an innovative way to get some help rebuilding his soil. A paper recycling plant located nearby agreed to bring their paper waste to the farm, rather than hauling it to the landfill. "I put 36,000 tons of wet paper on this farm," Joel remarks. "I prevented 1400 semi-truckloads of paper from going into a landfill, while also benefiting the soil by providing organic matter. I got a weed suppressive mulch, and they paid me! It was good for everyone." Ultimately, 215 acres of land were covered with two inches of paper mulch. Joel doesn't need scientific proof to believe that it was the right thing to do. He explains, "It was cool and moist under the paper; it was hot and dry where there was no paper. The grass was green growing up through the paper, and brown where there was no paper."

Joel's neighbors have been tracking his progress with some interest. He says, "At first their reaction was, 'This guy is a total nutcase. He's going to lose the farm and we're just waiting so we can buy it

up from him.'" He continues, "Then there was a period of more tolerance within the neighborhood. And now, I have had several guys tell me, 'You know what, people around here are not saying Joel's such a nutcase anymore.' They are actually looking at what I am doing."

This was a good year for people to be watching Joel. Thundering Hooves sold out of beef early, and Joel is confident that demand for hormone- and antibiotic-free, vegetarian-fed, custom-slaughtered beef will not diminish anytime soon. He comments, "In order for this thing to really be sustainable for our family, the employees, the infrastructure that needs to happen, things have to ramp up some more. What we have been doing with 160 cattle, 7000 chickens, 2000 turkeys, and a few dozen sheep and goats—is really only done on 40 acres of the 400. So we have ten times the biologic potential."

Joel's ready laugh softens his messianic enthusiasm, but there is no doubt he has been fully converted. He says, "I know what I am on to here is right, whether you are talking soil biology, whether you are talking about community, the farming environment, or social and environmental issues. I immersed myself and educated myself and had all these other failings, and learned from the process. I never make the same mistake twice, but I always make new ones. I am a big failure, but that's because I know what it takes to succeed." Joel concludes, "This farm has grown and evolved. You can't poke a flower back into its bud. It's out there and it will never go back—that's Thundering Hooves."

WILLOWS INN, NETTLES FARM, & LUMMI REEFNETTERS

Riley Starks and Judy Olsen didn't know when they jokingly referred to their new farm on Lummi Island as "Nettles" that the name would stick. When they first saw the parcel, just uphill from a venerable island institution, The Willows Inn, it was thickly forested, and stinging nettles were growing everywhere. Riley explains, "The Willows Inn is on nearly every map of Washington State, so we referred to our farm as The Nettles—the poor cousin of The Willows. It's a joke that kind of backfired on us."

On Nettles Farm and in the nearby waters, Riley and Judy have devoted themselves to producing food of the highest quality. More recently, they have expanded their enterprise to include the

historic Willows Inn, where guests can sample the bountiful fare grown just up the hill, and enjoy a spectacular view of the other San Juan Islands.

Riley was born due south, on the other side of Puget Sound, in Port Townsend but spent part of his childhood in France. His approach to food is distinctly place-based and European, as embodied by the Slow Food movement. "A lot of people say our pasta is the best pasta this side of Italy," says Riley. "That is sort of the ideal we aspire to. We can create that right here, we don't have to go to Europe. We can take from our surrounding environs and produce real food, and if you go somewhere else it is entirely different—and it should be. That's our mantra. Eat from your own landscape."

Riley acknowledges that the impetus to farm was driven more by emotion than a business plan. "We started out of desperation—not having access to decent vegetables," he explains. He and Judy cleared part of the forest and nettles on their land, built a house using some of the trees they had felled, and started to cultivate the soil in the protected clearing they had created. Riley has been a commercial fisherman all his life, and Judy worked as a cardiac nurse. "We started slowly," Riley explains. "Neither one of us quit our jobs, and we only farmed four acres."

Proving that tomatoes could be grown in the state of Washington without added heat, they showed up to the newly started Bellingham Farmers' Market with 18 tomato varieties of all colors. "It really floored people," Riley remembers. They invested in a commercial

kitchen to wash and package greens, but for much of the year—the non-greens season—the kitchen remained vacant.

When the opportunity arose to purchase pasta-making equipment, Riley and Judy pounced on it with the intent of making fresh pasta. It turned out to be more than simply a use for the kitchen space. They realized everything on the farm could be funneled into the pasta, and the pasta equipment was top of the line. It was a classic Italian design with an old, bronze die that roughened the edge of each noodle, enabling it to accept sauce. Their certified organic pasta business became the moneymaker. "The local co-op gave us our real

start," recalls Riley, "then we went to the high-end markets in Seattle and Tacoma." Still it took seven years and unflagging optimism for the pasta business to become profitable.

In Nettles Farm ventures, one thing often leads to another. Once they created the pasta, they recognized that eggs are the single most important ingredient in pasta, the one that makes all the difference. "You can use good flour, but if you use bad eggs your pasta stinks," says Riley. "If you use bad flour and good eggs, however, you have a decent pasta." Appreciating the need for organic eggs, Riley and Judy bought their first batch of chickens and built a chicken coop.

Before long, eggs were part of the Nettles Farm offerings, and organic chickens soon followed. "I love chickens, and I happen to believe that the chickens in this country are a poor excuse for food," contends Riley. "I knew that we could do better." Nettles Farm chickens are grazed using portable coops and slaughtered on site, by Riley, in a USDA-approved facility he built. Given their success, Riley and Judy finally quit their jobs and, as Riley puts it, "gave ourselves over to doing the farm."

Around this time, farm-raised salmon flooded the market, and Riley's commercial fishing work became less profitable. He joined in a Lummi Island tradition and bought a reefnet in 1992. Reefnetting is the oldest net fishing technique known, originally practiced by the native peoples of Puget Sound. Today it only occurs off the west coast of Lummi Island. In practice, reefnet fishing looks like a high wire circus act on water, with two stationary boats strung together, nets hung like hammocks between them, and a precarious-looking platform high above each boat with barely room for one person to stand to look out for the salmon. It's a passive fishery—as the armchairs on the decks of the boats attest—that relies on patience, currents, and underwater rope tunnels that look like kelp to attract the salmon.

Riley readily admits its inefficiency, but it works for them because they occupy an upper tier market niche and don't need to catch a lot of fish to make it profitable. Reefnet salmon is higher in omega-3 fats because they are caught so far from their home streams in British Columbia's Fraser River system. The gentle handling and processing techniques used in this type of fishing mean there is virtually no bycatch. "The salmon tastes better because it is bled live in sea water; this is the only fishery where that is possible," Riley explains. He is having success selling Reefnet Salmon exclusively to Metropolitan Markets of Seattle and directly to neighborhood buying clubs across the country.

Two years ago, Riley and Judy decided to shift their focus from producing high-end retail food to presenting model meals directly to consumers. The historic Willows Inn provided that opportunity. "We bought the Willows Inn because we wanted to take the connection one step further, right to the people on their plate," explains Riley. "We could produce the food and send it to retail markets, and that was pretty exciting, but it wasn't enough. We wanted to go further and have all of that goodness right on the plate. Then you are one-to-one with your customer and you can really do some education." To keep up with this transition, Riley and Judy sold the pasta business and downscaled their farm production to solely meet the needs of the Willows Inn's two restaurants and pub. Nettles Farm Pasta and Nettles Farm Eggs are now independently owned but are still produced on the farm.

Riley describes the philosophy that has guided all of their work: "We can do better than the choices that have been given to us—not just by talking or trying to get other people to do other things, but by doing better and showing people that it can be better. To have a model that works—to me that gives hope." Riley contends that part of their success is due to their ability to charge high prices for their products. He insists, "You have got to be willing to charge, and somehow it's got to be okay, otherwise you are stuck with the same model that everyone else has: the corporate model—the cheapest inputs, the lousiest labor, and the most mediocre product you can produce—and that's what we end up with as food."

"Well not us," Riley is quick to add with a grin, "we eat pretty well!"

RESTORATION

As human settlement and development have altered Washington State's landscape, the scarcity of natural resources and decline of biodiversity has become increasingly evident. But some people are working to reverse this trend by combining human ingenuity with new information and tools to restore natural habitats. Frequently, the key ingredient to success in these efforts is persistence.

In the Northwest, endangered species, particularly salmon, are commonly the driver for habitat restoration, and the region's native peoples are often the voice of the fish. With technical help and expertise from local stakeholders including tribes, environmental organizations, government agencies, and private enterprise, landowners are learning how landscapes can be improved to better meet the needs of multiple species and interests.

Where Washington's settlers applied themselves to the task of taming the landscape to increase human productivity and comfort, the resource managers in these stories are working to accommodate a wider array of species, ameliorating some of the impacts of past generations.

GOLDSBOROUGH DAM REMOVAL

Debates about the merits and costs of removing dams have been in and out of the national spotlight for years. While most dam removal efforts have been bogged down in logistics and opposition, on Goldsborough Creek in south Puget Sound, a tribe and a timber company have teamed up with state and federal agencies to take out an old dam and bring salmon back upriver.

Goldsborough Creek drains 80 square miles of forestland on the Olympic Peninsula. The creek had been dammed since 1885 when a railroad company built a log storage pond on the site. In 1921, the dam was used to generate power for the growing community of Shelton. After a flood wiped out the dam, it was rebuilt in 1932 to provide power and to divert water for a forest products operation. Simpson Timber Company acquired the dam in the 1950s, but ceased drawing power from it in 1986.

rendering the dam's fish ladder impassable. The dam became utterly useless to Simpson in 1996 when flooding destroyed the water diversion pipeline.

When it was rebuilt in 1932, the dam was 100 feet long and just 14 feet high. But with water flowing over it at an average rate of 400 cubic feet per second, the streambed below the dam was soon scoured out, adding another 20 feet to the dam's height and

The Squaxin Island Tribe—also known as People of the Water—have long been stewards of southern Puget Sound's salmon. They had advocated for the removal of Goldsborough Dam for over 20 years, knowing that upstream there is over 25 miles of main stem and tributary habitat. According to Jeff Dickison, a tribal biologist, "I've worked on this project for 15 years, and it was already a file when I was hired."

Jeff remembers when the tone of discussion about dam removal changed. "The one meeting I'll never forget was with a Simpson accountant," Jeff recalls. "We were having this whole discussion about the dam and the costs of repairing some damage and the risks associated with it. The accountant was busily writing down numbers and taking notes, and at the end of the meeting he said, 'It appears to me that this facility is probably more of a liability than an asset.' That changed the whole picture then and there."

Not that dam removal was a done deal once Simpson Timber Company and the Tribe were in agreement. The next step was to gather the five million dollars needed to fund the project. The Washington Department of Fish and Wildlife and Simpson both put $1.1 million toward the project, and the state also agreed to take ownership of the dam, assuming liability for its removal. The Tribe played the role of matchmaker, which meant a lot of lobbying in Washington DC.

Then came the lucky part. A law passed in 1996 authorized federal funding for fish restoration projects. Jeff explains, "The way it works in Congress is that they pass lots of laws with authorizations and very few appropriations to back them up. We realized the law was there, the authorization was there, and we needed to get the appropriation." With the joint effort of Simpson, the Squaxin Island Tribe, and Congressman Norm Dicks, the federal dollars were appropriated through the Army Corps of Engineers. The Goldsborough project was one of the first to take advantage of the new funding source.

Removing the dam and reestablishing the streambed was a significant piece of engineering. "Because of erosion of the channel downstream, it was past the point of just taking out the dam," explains Jeff. "If you had blown the dam up and let it try to restore itself, there would still have been an impassable barrier and the fish still wouldn't have been able to get upstream."

Work began early in the summer of 2001, when stream flows were low. First the stream was diverted into 2000 feet of four-foot diameter pipe. Removing the old wooden dam was relatively easy; the more difficult work lay in creating a gently sloped streambed to replace the 35-foot drop the dam had created. The sediment that had collected behind the dam was used to build up the downstream stretch. Thirty-one concrete weirs were installed across the streambed, creating a series of low steps that stabilized the rebuilt slope. Then, numerous boulders and large tree stumps were scattered between the concrete weirs to improve fish habitat. Jeff explains, "The large woody debris creates more turbulence, particularly at higher flows, which scours underneath the logs leading to deeper pools and shaded cover for fish."

Jeff acknowledges that the work of rebuilding the streambed didn't always look like people's idea of restoration. "At times it looked like a highway project," he recalls. "It was this dewatered stream that they were running big equipment up and down, moving all kinds of earth." Typically, large equipment is not allowed anywhere near salmon-bearing streams, and vegetative buffers are required, but for this project, many streamside trees had to be removed. Once reconstruction was complete, alders and willows were planted along the stream and slowly they are beginning to restore the shade canopy.

Water was returned to the restored streambed in November of 2001 and chum salmon were spotted above the old dam site that fall. The next year, the Tribe counted more than 15,000 chum migrating down through the newly restored habitat. There is a lot of potential for coho production in the creek, which is the species the partnership is hoping to significantly improve. "It is definitely highly engineered," says Jeff. Some people still express concern about that, as well as about the nature of the habitat conditions in this reach. "Whenever you do something like this there are always trade-offs, and the real trade-off here was 25 miles of habitat upstream." Jeff continues, "To me this is about fish passage—look at it as a fish ladder, as whatever you want—it's about getting the fish past this point, and there is great habitat upstream."

Part of Jeff's optimism is due to the commitment Simpson Timber Company made to protect the upstream habitat now accessible to

salmon. Simpson has since split into two companies, and one of those, the Green Diamond Resource Company, now manages the upstream forestland. In a move that put them in front of the industry in terms of environmental stewardship, the company developed a Habitat Conservation Plan designed to prevent future endangered species listings and degradation of the Goldsborough Creek ecosystem. Jeff explains, "The Plan provides really stable land management strategies well out into the future. So for this vast watershed and 25 miles of stream, you have a pretty good idea of what it's going to look like for quite some time."

HEALING HOOVES LLC

Knapweed, tansy, leafy spurge, blackberries, and English ivy are just some of the weeds and invasive species that give land managers headaches. Craig Madsen is offering a picturesque, poison-free solution: his fetching herd of goats and sheep will forage freely on the stuff you don't want, and leave the more desirable greenery behind.

Of course, it's not quite that simple, as Craig would be quick to tell you. But given weeds at the right stage of growth—or palatability—some good fencing, and an intensive rotational grazing approach, goats and sheep can be a very effective tool for weed control. There is an art to this science and as a shepherd Craig does much more than just move his "flerd" (flock of sheep + herd of goats) from one place to the next and keep predators at bay. He has 15 years of professional range management work to draw on, and a lifetime of ranching experience.

grazing service because people don't think they should have to pay for grazing. I call it weed control."

Craig had grown up with cattle, so he wasn't sure he'd get along with the sprightlier livestock. "I bought a few goats three years ago to see if I could tolerate them and they could tolerate me," he says. "They worked out fine; they are fun animals." Craig and his wife, Sue Lani, launched Healing Hooves in 2002. Their first project was for the US Fish and Wildlife Service, at a fish hatchery in Leavenworth that was interested in alternatives to spraying weeds with herbicides. Since then, Craig has worked with some private individuals in the Methow Valley, and has even unleashed the flerd on invasive blackberries in urban Seattle. Most of the time he works on projects that are 20 acres or less.

Craig helps clients to develop long-term ecosystem and herd management plans, but his main focus is vegetative management. "I provide a weed control service, a service you pay for just like spraying or mowing," he explains. "It is just another tool you can use depending on the objectives for the site. I usually don't call it a

"Now I have 250 head, mostly goats and a few sheep," reports Craig. "The sheep are a shedding variety, so I don't have to shear them; they are more of a meat sheep. All the male goats go to the meat market in October, and I keep the does over the winter and they kid in the spring." Craig has tried selling cuts of goat meat

directly to consumers, but it wasn't cost effective; he now sells all his meat goats live to a couple of buyers.

"Most people think of weeds as a problem, and they can be, but for certain types of animals they are forage," explains Craig. "As for a goat's palate, broadleaf plants and shrubs are tastier, more nutritious fare than grasses. There is more concentrated protein in shrubs than grass and since goats have smaller stomachs than cows, they select for the higher nutritional value feed." Since goats prefer to eat higher up on a plant and sheep tend to eat lower down, the animals work well in tandem.

"There are all kinds of potential uses for goats," comments Craig, "it's just a question of fitting them into a situation where they will be effective for weed control or reducing fuel loads." Goats do especially well in sensitive areas, like those near wetlands or streams. They're also uniquely suited to difficult terrain. For instance, they are agile on steep slopes, and willing to jump up to reach hard-to-get foliage. And, they don't mind thorns; blackberries are some of their favorite treats.

Craig and his herd are currently working on a three-year USDA-sponsored project to evaluate goat weed control potential on ranches in the federal wetland reserve program. The USDA has been using chemical sprays on Russian olive, an invasive species that spreads quite rapidly in wetland areas. The plant is infamous for its nasty thorns, which are even capable of flattening tires. "We are evaluating how effective goats are for managing Russian olive compared to chemical treatments," Craig explains.

With a guard dog on patrol, in some ways the life of a modern shepherd looks much as it always has—with a few extra tools. Craig illustrates, "When I take animals on a project, I put up portable electric netting; I fence the area where the weeds are and

put them on a half acre or an acre at a time. Then I move them once a day or so." Much like a traditional shepherd, Craig is mostly on the move. For six months of the year, he lives out of the sleeper cab of the truck he uses to haul the "flerd" from one job to the next. While many couldn't imagine such a lifestyle, it suits Craig well. "I enjoy being outdoors," he explains. "I move the animals every couple of days and I am always checking on them to make sure they are where they are supposed to be and that they are not being chased by something. For work like this, you've got to enjoy being by yourself, and being outdoors and camping."

During the cold wet months, finding food and shelter for 250 animals is a costly challenge, especially since goats do not like being wet. Since Craig and Sue Lani own just a couple of acres, they rent other pasture in the winter and coordinate with neighbors, swapping pasture for weed control. The other challenge during the winter months for Craig is lining up a continuous flow of work for the coming season. While marketing is not Craig's favorite part of the job, he realizes that building relationships with landowners is what will continue to make his business successful.

Craig doesn't claim that controlling weeds with animals is a quick fix. "If just weeds are there," he explains, "you have to work with the landowner to establish something to compete. It's not a one-time shot; weeds are very well adapted to producing seed. If the plant has been there several years, there is a seed source in the ground that will continue to produce new plants for many years. You have to take a long-term approach to weed control. You can't just hit it once and be done." But as any landowner can tell you, there are no easy fixes for invasive weeds, and most people find that having a herd of weed-eating goats on their land is a lot more fun than spraying chemicals.

HINES MARSH RESTORATION

Hines Marsh is a remote wetland on the very edge of Washington that until recently attracted little attention. Even savvy birders—who flock to the far end of Long Beach Peninsula—drive right past the forested edge of the marsh without even knowing it's there. But recent decades have been anything but peaceful in the marsh. The triumphs and tragedies that have played out here contain enough drama and intrigue to fill a hefty novel.

The lead protagonist is appropriately beautiful and mysterious. The Trumpeter Swan is the world's largest waterfowl, with dramatic all-white plumage and a black bill. In the early 1900s, Trumpeter Swans were hunted to near extinction; in 1932, according to the Audubon Society, only some 70 birds were known to exist, and none of these were on the west coast. Then in 1950, a flock of Trumpeter Swans, 80 strong, from central Alaska was discovered wintering on Hines Marsh.

According to Martha Jordan of The Trumpeter Swan Society, "Hines Marsh was, at one time, a beautiful, pristine, interdunal wetland system, which means that it lies between some major sand dunes." The Long Beach Peninsula is a 27-mile-long skinny stretch of sand that separates the Pacific Ocean and Willapa Bay; sand dunes run north-south almost the whole length of the peninsula, which is attached to the mainland at the south end. Historically, the marsh was 700 acres of freshwater on the far north end of the peninsula, making it the largest interdunal wetland system on the Pacific Coast. Its remote location and wet, swampy terrain meant it was relatively untouched by developers

until 1962. Then developers—with visions of houses on canals and an amusement park—ditched and drained the marsh. The swans did not return that winter, and the developer's money dried up along with the marsh. Martha laughs, "All we can say is, thank God for bankruptcy!" Martha was introduced to the remnants of Hines Marsh, some 20 years after it had been drained and abandoned, by a local oysterman who had a dream of restoring the marsh.

"The oystermen are very vocal in the community," Martha reports. "These are good people and they really care deeply about the health of the bay." Martha believes that the reason Willapa Bay is as pristine as it is—and it is the most pristine bay in America today—is because the oyster industry is there. "They are the defenders," Martha explains. "They have a vested interest in it. The marsh directly affects the quality of water in Willapa Bay."

The Willapa Bay oystermen had found the advocate they needed in Martha, an independent wildlife biologist based out of Everett. As Martha tells it, she became the leading swan expert in the Northwest by default: "I never wanted to work with swans in the first place—they picked me and insisted that I work with them." Martha and the swans and the oystermen changed the fate of Hines Marsh in ways they could never have anticipated.

The first step towards marsh recovery occurred in 1984, when mitigation for a nearby road led to reconstruction of breached dunes within the marsh, effectively restoring the natural water flow. The county needed a nonprofit organization to take ownership of the two small dune parcels, and since Martha was already there working with The Trumpeter Swan Society, she saw the opportunity for the Society to take ownership of the restored dunes.

Over the years, the dunes sat there innocuously doing their job, but it was a controversial topic among some politicians in the county. Some blamed the dunes for flooding a road to the south. Martha recalls, "They did some political maneuvering and they tried to get the dunes removed, but nothing went anywhere because legally they couldn't. Wetlands are protected!"

In 1989 Martha got wind that 140 acres were coming up for sale at the north end of the peninsula, including the outflow of the entire marsh. She had three days to get the down payment of $7000 together, which she did with just five phone calls to oystermen and The Trumpeter Swan Society donors. Two years later the land was purchased by the Washington State Parks Department to assure its long-term preservation, and Martha was ready to take a breather. "I thought, okay, phew!" she remembers. "I had other things to do. It's four-and-a-half hours away from where I live. It's not like it's across the street." A few more years passed quietly in the marsh, but Martha was destined to get a lot more familiar with that commute to the Long Beach Peninsula.

The next phase of marsh restoration had an unlikely and an unwilling benefactor. A Pacific County Planning Commissioner grew frustrated with the "problem" of water in the marsh. He decided to take matters into his own hands. In early 1998, he hired two men to take his backhoe out, on the sly, to remove the marsh's east dune. Martha remembers discovering the marsh dry on a Sunday in January, and hiking in to find the east dune gone. "Water was just gushing out, trees were down everywhere," she recalls. "It was horrible, as if the marsh's carotid artery had been cut. It was the worst thing that could possibly happen in the whole system."

She then drove over to oysterman Dick Sheldon's house. "I knew what it meant to the oystermen, because we had several thousand acre feet of fresh water and sediment draining out in January, the prime oyster fattening time on their prime beds," she explains. "It was Sunday afternoon in the middle of the Super Bowl. I walk in and I said, 'Dick, we have a problem.' And Dick says, 'It's the Super Bowl. Can't it wait?' I said, 'Dick! They've taken the east dune.' He said, 'No way!' And I said, 'They've taken the dune. It's gone.'"

The stealth removal of the dune proved to be the largest hydraulic permit violation and one of the biggest shoreline permit violations in Washington State history. The truth of what had happened very slowly emerged thanks in no small part to the diligent efforts of two Washington State Fish and Wildlife law enforcement officers. Martha was in the midst of a very ugly local battle. "It was the fight of a lifetime," she admits. "I wasn't sure I was capable of it—but what these people were doing was so wrong. Because of personal threats, it was suggested I carry a cell phone, never go anywhere alone, and consider a concealed weapon." Local county officials dragged their feet on issuing permits for replacing the dune, but Martha persevered.

"In the end we were given full restitution," Martha reports. "The County Planning Commissioner resigned but decided he would do most anything to avoid paying us. So we liened his property and then exercised our lien. He had to sell his 240 acres and went into bankruptcy and we were paid within a year." That infusion of cash meant that The Trumpeter Swan Society was finally in a position to take the next big step toward restoring the marsh and realizing the dream of bringing the swans back.

Over the 20 years the marsh had been drained, a lot of willows had grown in. When the marsh was reflooded with water when the dunes were restored, the willows died. Martha recalls, "There was this sea of standing dead woody debris. You couldn't even canoe in it, it was so thick." So the logical next step was to create open water in the marsh, to repair the aging water control structure, and put in a fish ladder. It was the ethereal dream, and more than $130,000 was needed to make it happen. With the settlement money and the notoriety generated by the dune destruction, the dream was in reach.

Resources and partners to restore Hines Marsh were no longer in short supply. "Hines Marsh went from obscurity to the spotlight with Columbia Land Trust, Washington State Parks and Recreation Commission, Ducks Unlimited, and the Audubon Society," Martha explains. With these and other partners, the hard, costly work of clearing the marsh and making needed repairs began. With hip waders and underwater chain saws, 40 acres at the north end of the marsh were cleared. Another 120 acres of marsh habitat was obtained and turned over to Washington State Parks.

In January 2003, Trumpeter Swans returned to the marsh for the first time in 40 years. "That first year, six swans came back to the marsh," Martha says with a smile, "and this last year we had thirteen." With the swans back, the future of Hines Marsh seems more secure. She reflects, "When people see the swans, they are just awestruck. The swans are the greatest ambassadors in the world—whether you are the most hardcore property rights person or the most hardcore animal rights person or somewhere in between—you love the swans. The marsh brings people together."

While the elegant birds have undeniable appeal, the restored marsh also provides habitat for other waterfowl, otter, bear, beaver, deer, elk, bobcat, and potentially coho salmon. In addition, the marsh provides significant flood control and water quality protection, acting as a filter for groundwater resources and runoff into Willapa Bay.

Martha and the others working to protect Hines Marsh are not resting on their laurels. There are still threats from adjacent landowners, and water rights to protect. "Our original concept was to have this be a wildlife area and to keep people out," Martha explains. "But the marsh needs to be shared. The more people that know about this place, the more eyes that are on it, the better it can be protected."

RESTORATION

JIMMYCOMELATELY CREEK RESTORATION

When a section the northern reach of Highway 101—the main arterial for Washington's Olympic Peninsula—was closed for 12 hours due to flooding in 1996, local, state, and tribal agencies were spurred to action. But instead of building culverts and flood control devices to further restrain Jimmycomelately Creek, they decided on a more radical plan: to allow the lower stretch of the creek to flow unimpeded back in its natural floodplain.

Jimmycomelately Creek—or "the Jimmy" as it is known—was moved uphill from its original location over 100 years ago by farmers looking to cultivate the fertile valley soil. Out of its floodplain, the creek cannot naturally dissipate during high flows, inconveniencing people and decimating salmon habitat. Once this project—the largest creek meander restoration effort in Washington—is completed, threatened summer chum salmon will once again have access to prime creek and restored estuary habitat.

"We like to call it an 'undevelopment' project," explains Sam Gibboney, project coordinator. "The goal is to pull back the human development that has occurred and give the creek a chance to move and do its thing naturally. That's a driving philosophy behind the restoration project; we are not really in the business of trying to create habitat so much as we are in the business of giving some room in form and function and letting the habitat recreate itself."

But before work could start, there were a few details to work out, particularly since the original floodplain at the mouth of the creek was pretty well filled in and built up, with two roads, a log yard, a

dance hall, and even an RV park where wetlands and the estuary used to be. The Jamestown S'Klallam Tribe had its eye on the log yard for over a decade because of their interest in salmon and shellfish harvests in the tidelands.

Sam has a contract with the Tribe, but she works for and with all of the project partners. "The property acquisition phase really took a lot of creativity," she explains. "The Tribe was able to purchase the old log yard, which is quite an extensive piece of property, with funds from the US Fish and Wildlife Service. The Washington Department of Fish and Wildlife partnered to purchase several other properties in the area including the old RV park." At first, negotiations did not go as well with the owner of the old dance hall, but the Washington Department of Transportation stepped in and was able to acquire that key piece of land.

"The Tribe has a dual role in this effort. They are a project partner and also an affected property owner. They have allowed the project to happen on property they could use for other purposes. The Tribal council believes strongly in this project and has been a driving force," explains Sam.

A high level of cooperation and coordination has been required to put the Jimmy back where it belongs. The Tribe's previous natural resource director used to have a saying in her office: "It's amazing how much you can get done when you don't worry about who gets the credit." Everybody—agencies and individuals—have kept their eye on the end goal of the project and that has kept things moving

forward. Sam contends that private landowner participation has contributed to the success of the project. "A key landowner is John McLaughlin, who allowed a conservation easement to be placed on his land," says Sam. "John also acts as a representative for the landowners and having him talk to his neighbors about ongoing issues rather than agency people has helped facilitate local cooperation."

The project includes removal of over 100 creosote pilings in the bay, along with a column of contaminated sediment around them, in the hope that tribal shellfish harvesting will eventually resume. In

addition to removing structures, septic systems, fill dirt, utilities, and roads, there is one significant piece of human engineering that is being built: a new Highway 101 bridge that will span the restored creek and its entire 100-foot floodplain.

Designing three-quarters of a mile of restored creek bed has been part science and part politics; an 1870 coastal survey has been a useful reference. "While there are a lot of biological principles that drive this project, it is also within the context of current land use," explains Sam. "A channel design team met once a month for close to two years, and looked at everything from location to function to how much wood was going to be placed in the channel, and came up with this design based on land ownership and willingness to cooperate."

Sam explains, "We are not putting the creek back into the exact place it was before. That would be impossible. But it fits within the landscape now, and we are adding a lot of sinuosity back into the channel to make it function better." Where the RV park stood, there was some question as to whether that area was historically salt marsh. However, the design team came to the conclusion that salt marsh is at such a high premium in the Puget Sound right now, that it was okay to depart slightly from a pure restoration philosophy to try to create as much salt marsh as possible because it is so important to not only salmon but to other species and habitat.

As project coordinator, Sam spends her time both in the office assuring the funding—currently from 18 different government sources—and permits are in place, and visiting the various project sites in hardhat and waders. "While the project was long in forming from conception through acquisition," Sam says, "once we actually got to design and construction we have had a very aggressive schedule." A lot of what drives that is the fish. Volunteers have been capturing some of the summer chum that return to the Jimmy each year and artificially spawning then releasing them, but everyone is anxious for the salmon to spawn naturally.

"Working on a landscape scale is really exciting," comments Sam. "Not just doing band aid fixes on different parts of the creek, but really looking at it on the broader scale of how this fits into Sequim Bay. One of the best things about this project is how much land will be held indefinitely in a natural state." This project is unique in that it links a restored creek bed to its estuary, taking riparian restoration to a grander scale which the Tribe hopes will serve as a model for other projects. For her part, Sam is looking forward to the moment when human undevelopment—with all the backhoes and trucks it requires—gives way to something else. "We are getting all the pieces into place—or actually out of place—so we can just let her rip."

NUQUA'LUM—
RESTORATION AT MAPLE CREEK

Gary Gehling was a developer that had a change of heart. In 1996, he and a partner bought 113 acres a few miles south of the Canadian border, where Maple Creek runs into the North Fork of the Nooksack. Their plan was to subdivide the property into 22 parcels and build houses. "It didn't take us long to realize it wasn't the best thing for the place," says Gary. His partner didn't quite see things that way, so Gary bought him out and then arranged for all but 25 acres to be transferred to a local land trust. Now Gary and his family enjoy daily wildlife sightings in their backyard, and they are finding alternate ways to sustain themselves on this land.

When the Gehlings bought the property, it was an old dairy in disrepair. Gary remembers, "The first thing we did was start cleaning the place up. There were some old barns that were falling in and just tons and tons of trash. A friend suggested that Gary contact the Lummi Nation about doing some restoration work on the creek and so he did. "They came and were very excited about doing a project up here. The streambed was nothing but reed canary grass," he recalls. "The grass forms a real dense mat and slows the velocity of the water and then starts silting in behind it."

Taking advantage of a cost share program with US Fish and Wildlife, Gary had the Lummi Nation come to do restoration work on the creek. He comments, "It's good that people can make a living doing this kind of work; the short-term gain is jobs, and over the long term there is a fisheries benefit." Gary and his family did a lot of the work too, preparing areas for planting and gathering trees and willow cuttings to be planted along the banks of the creek.

The original restoration project involved a 200-foot buffer strip along the creek and a 300-foot buffer along the river. While the river bank had a lot of trees, they were all hardwoods and there was no seed source for conifers. To address this, they planted conifers in the understory to speed up succession. "I think in that first year we planted 12,000 trees," reports Gary. There have been more than 20,000 trees planted over the years. The hope is once the canopy grows in, it will shade out all the non-natives like reed canary grass. "I don't think canary grass is going to go away," says Gary, "but at least we can give some native species a chance to come in."

Gary is clearly smitten by the wildlife that takes advantage of the restored Maple Creek. Spawning salmon are a main attraction for visitors and wildlife alike. "The kings come up in July; sockeyes come in shortly thereafter; cohos come in the fall," Gary explains. "Pinks run every other year at the end of August; chums are in November and are around until Christmas. There are local rainbow trout and the steelhead come up when they feel like it. The riparian areas are like a highway for the wildlife, like a magnet. We see signs of bear just about every day. We see bobcat, elk, and cougar tracks along the river. There are otters, especially when the pink salmon are in. Otters will grab a salmon and take one bite where the gills are—which is a real fatty spot—and just toss it aside. You'll see a pile of 20 or 30 fish with one bite out of them." There are also rabbits, coyotes, and pretty much everything that is native locally. In the wintertime Gary and his family see lots of eagles. "I have counted 17 on the bridge when there are fish in there," he says. Canadian geese and blue heron sightings are a daily thing in season. Gary notes, "In the snow we watch bobcats for hours at a time, pouncing on voles in the field."

The financial side of Gary's revised plan for the property has not yet been as rewarding as the wildlife viewing. "We do intend to develop the place," he explains, "and make it habitat for people and wildlife too, but it will be low impact where people are here on the weekend and go away the rest of the time." Thanks to a simple web page, couples from across the country come here in the summer to exchange their wedding vows in a unique and spectacular outdoor setting. Local hotels, stores, musicians, and caterers have enjoyed the additional business the weddings generate.

Gary and his wife Ellen plan to create some additional spaces so weddings and other events can occur year round. They are trying to get the capital together so they can have a building for workshops and conferences, and they'd like to do environmental education seminars, too. "I could go out and borrow money," remarks Gary, "but I don't want to. I think we are better off running this as a blue sky business with low overhead, because the overhead is what ties you into to having to deal with short-term profits. Everybody wants their short-term profits, including the banks. So you kind of have to be creative and come up with your own money."

Gary realizes his plans will not bring the financial return that a housing development would have. But over the long term, as property with intact wildlife habitat becomes more and more scarce, he believes there may be opportunities to increase revenue from the land. "But I am not really looking at it in those terms," he explains. "This is our home now; we are looking at how we can make a living without having to leave here."

"We like providing a place where people can come out and enjoy wildlife habitat without being destructive about it," he continues. "They can come out here and have a party and then they can go home. Everybody gets to see salmon spawning—they think that's great, let's save salmon! Education happens here without having to get all preachy about it—people see what federal dollars for salmon restoration are doing."

Gary has given his place the name Nuqua'lum. It is the Salish name for the North Fork of the Nooksack River. "I found it in a local history book," explains Gary. "I couldn't say it the way they said it, it wasn't a written language and it has been changed over the years. We had one of the Lummi elders out here for a ceremony and he said 'that is close enough.' They were honored that we would use their traditional name. I like to honor those who came before us."

"In this dollar driven economy, short-term profits rule; any sustainable economy is more long term," Gary says. "But you've got to do what you believe in. What is your place on the planet? What are you leaving behind? I am proud of houses I have built, but if they were here I don't think I'd be proud. This is a special place."

RESTORATION

SKYKOMISH RIVER BANK AND HASKELL SLOUGH RESTORATION

When a good bit of Dale Reiner's 300-acre cattle farm was inundated by flooding in 1990, he didn't realize that this disaster would shift the focus of his work on the land that has been in his family for generations. By reaching out to cooperate with unusual allies, Dale has forged partnerships that are bringing new hope to those who value rural landscapes and salmon habitat.

The Reiner family has worked this parcel of land, at a bend in the Skykomish River, for four generations. "My great grandfather homesteaded here in 1873, and my grandfather was the first white child born in this area," says Dale. "People like my grandfather and great grandfather filled in these backwaters and sloughs and oxbows, doing what every farmer did to create additional agricultural land."

When a railroad and highway were built across the river from Dale's land, a rock wall buttress was installed that effectively cut off the north portion of the floodplain. Thus during very high flows, the Skykomish River spills over onto Dale's property. With the flooding in 1990, the river spread out over 100 acres of Dale's cattle pasture, taking out fences and leaving behind sand, rock, and sediment. The county estimated that another flood could damage nearby highways and said they would build a berm to bolster the riverbank. So Dale went ahead and invested $365,000 in rebuilding fences and cleaning up after the flood, but the berm didn't get built, and in 1995 there was more flooding.

A friend suggested that Dale enlist the help of John Sayre, director of Northwest Chinook Recovery. Dale remembers, "I figured if I had the environmental community working with me, I'd get a lot farther with permitting and getting things started." John had been in the fish restoration business for a long time. He came out and looked at the slough and the series of ponds and said, "Wow, this is fantastic."

The eleven ponds strung along Haskell Slough provided excellent slow-water rearing habitat for endangered salmon during high

flows, but they lacked an outlet back into the river, so fish were likely to end up stranded once the water receded. "The Skykomish River is the second most productive wild salmon river in Puget Sound," says John. "You've got about 20 percent of the remaining wild chinook and 50 percent of the remaining wild coho here, and strong runs of all the other species."

John saw the potential to reconnect three miles of slough with the main channel of the river, and he had the ability to create a public-private partnership to make it happen. Northwest Chinook Recovery worked with Dale, his neighbors, and tribal, state and federal agencies to excavate 7000 feet of the channel, linking the string of ponds back up with the main channel.

The Haskell Slough project was widely celebrated by politicians and local leaders of all stripes—even then presidential candidate, George W. Bush, paid a visit. It was voluntary and it was effective. "By June 1999, we had approximately 3000 coho salmon smolts moving out of the slough on their way to the ocean," says John. "The next year we had 6000." Dale adds, "All the work was done through cooperation, no regulation—tribes, landowners, and environmentalists working toward a common goal."

The attention garnered by the voluntary restoration of Haskell Slough finally helped leverage some assistance for Dale with his flooding problem. Now that state and federal agencies had invested $700,000 in salmon habitat restoration in the slough, there was more support for protecting that investment. But it still wasn't easy. Some environmental groups were concerned about the precedent it would set to allow interference with the "natural" flow of the river. Says Dale, "They didn't want it to happen and they figured if they could stop this project they could stop any of them. We felt just as strongly that we could set a precedent that it wasn't just fish against farms or fish against people—something could be done to the benefit of both fish and people."

"What started as an incredibly controversial project is now the state of the art for how to restore a riverbank in a fish-friendly manner," recalls John. People come from all over the state and outside to tour the site. The innovative flood control structure is made up of two long rows of deeply-sunk pilings interwoven with large stumps and logs. Another row of thick cottonwood trunk segments, 1600 in all, is sprouting behind the weave of dead wood to provide a supplemental living barrier. The project is designed to slow the flooding river waters, protecting Haskell Slough and the homes, farms, and roads downstream. "If it works correctly," explains Dale, "eventually the bank will be built up to eight feet tall or better of gravel and sand. It will be a solid bank with trees growing in it."

In the course of these projects, Dale's focus shifted, and now in addition to the Angus cattle and Christmas trees he grows, he is attentive to the salmon. Nearly 150 acres of his land—fully half of his acreage—is dedicated to salmon habitat rather than farming. He seems unperturbed by the gravel, sand, and stumps that now dominate what was once his pastureland.

Through a federally-funded program that supports agricultural landowners who voluntarily plant and maintain vegetative cover adjacent to targeted rivers, Dale cleared 81 acres of invasive species and replanted with trees that will create a canopied buffer zone next to the river where his cattle once roamed. He maintains a 200-foot buffer along the river, and additional setbacks along the slough. Dale now has just 30 head of cattle, where he once had 430.

John sees this project as a model of innovative cooperation for the region. He says, "The agricultural community in the entire state was watching to see if Dale was going to get a fair shake. Now a number of other landowners are willing to consider allowing their land to be used for salmon habitat restoration purposes."

Dale is not resting on his laurels. His involvement in salmon restoration brought him to the realization that local tribes and farmers have a lot more in common than he previously thought. "On one side of the high-water mark I'm raising beef cattle and on the other side the Tulalip Tribe is raising salmon," he remarks. "We have many things in common: independent farmers, independent fishermen. We started looking at these similarities and realized, we shouldn't be fighting over issues, we should be working together." So with the help of Herman Williams, who was then chairman of the Tribe, they took things a step further to see what tribes could do for the farmers to help sustain them, and what the farmers could do to minimize waste and ensure clean water for the tribal fisheries. They arrived at the idea to do a biogas energy project to collect dairy waste from some of the neighboring farmers. This joint alternative energy project is close to fruition, and Dale anticipates more collaboration will follow.

John sees an important precedent being set: "If we're going to save these valleys, it is going to take a much bigger approach than just trying to do salmon restoration projects on individual pieces of land. You've got to try to save farmland and open space, otherwise you don't have any habitat areas to work with. So we try and build alliances with environmental groups and all the different governmental entities and the farmers and tribes. It's been an evolution in all of us."

TUCANNON RIVER RESTORATION— DON HOWARD

Lewis and Clark enjoyed a lunch break overlooking the Tucannon River in May of 1806. Their journals describe willows and cottonwoods growing along a pebbled creek with low banks. Despite early settlers' efforts to shape and control the flow of the river, it looks much as it did 200 years ago.

The Tucannon River originates close to the Oregon and Idaho borders, in the lower eastern corner of Washington. It is a key tributary of the Snake River, and home to no less than four endangered salmon species: bull trout, steelhead, and spring and fall chinook. Of the problems that often affect salmon-bearing streams—lack of habitat and insufficient amounts of muddy, warm water—only high temperatures are a consistent problem here. The relatively good condition of the river is due, at least in part, to the fact that its entire course is sparsely populated, and that many who do live here have worked hard to be good neighbors to the fish.

Don Howard is one of the few folks who has the Tucannon in his backyard, and his efforts are one reason why this river is in such good shape. About four miles of the Tucannon run through Don's 4000-acre cattle and alfalfa operation. Don's great-grandfather

settled here just 80 years after the Corps of Discovery passed through. Like Lewis and Clark, Don doesn't call the Tucannon a river. "It was always a creek to me," he says. "Most people come in here and call it a river, but I call it a creek."

Don's priorities are no different than those of his ancestors. "I want to save the stream bank; that is my number one issue," he asserts. "It is human nature to save your property from channel migration—my granddad did it, my dad did it." But Don has had to contend with a much more complex regulatory environment in addressing this priority. Since 1992, when the salmon runs in his backyard were listed as endangered, Don has worked collaboratively with local, state, tribal, and federal agencies to meet his objectives as a landowner, while also providing habitat and shelter for native fish.

Don is optimistic about human ability to come up with solutions to stabilize stream banks while providing for fish and wildlife. His stretch of the Tucannon runs pretty straight and fast. "Around 1993, they said we needed more pools," he explains. "I said, 'Well, let's build them!' They laughed at me the first time I said that." But Don gave it a try, and found that configuring rock and woody debris in the stream to scour out pools creates habitat and slows the water,

protecting the integrity of the stream banks. Don is quick to point out the experimental nature of these efforts: "Some are good and some are not very good, and the next year it changes. You never know what you are going to end up with."

Don has also experimented with planting trees in problem stream banks. The old method of shoring up a failing stream bank relied on rip-rap, which is usually a stacked wall of large rocks or uniform cement blocks. But it is difficult for plants to take root through rocks and cement, so stream shading and habitat complexity are diminished. Don describes an alternative approach: "I had a raw bank and it was cutting in really bad. I talked to the Department of Fish and Wildlife and said I'd like to do something here, so they came in and helped; it was a cooperative deal. We smoothed the bank up, put down palm matting and planted trees. Looking at it today, it's almost a little jungle there. Now we have trees to collect sediment. Maybe sediment will come in and kill the whole thing—then we'll go in and plant again." Don admits that they have had failures, but insists that if you don't try something, you will never gain anything.

Some of Don's management activities were not intended to protect salmon habitat, but they have had that effect all the same. "When my dad was still alive, my brother and I built a sediment basin to catch what came out of a canyon up here," he explains. "Lots of times the stream would be clear until it got to that canyon—it's right where snow drifts pretty heavy in the winter—then we'd get a thawing rain onto frozen ground. I was thinking we needed to catch some of that silt. I wasn't thinking about the creek at the time, but it made a big difference—we caught a lot of sediment there." Don, along with the vast majority of landowners in Columbia County, has also switched to no-till farming, reducing stream sedimentation.

Instead of plowing fields in the fall and leaving them vulnerable to erosion all winter, new seeds are punched into the matted remains of last year's crop every spring. The roots and stubble left over the winter hold the soil in place.

Improving endangered salmon habitat requires many more consultations and partnerships than most independent farmers can happily tolerate. Don has partnered with the Nez Perce Tribe to fence his cows out of the river. He has had fly-fishing clubs from the Tri-Cities come help him build rock structures to put in the stream. He uses only half of the water that he was originally entitled to, and the rest is held in trust by the state.

Don also allows some of his land to be managed as a riparian buffer as part of the federal Conservation Reserve Enhancement Program. He is currently working with the state's Department of Ecology to change his irrigation system to increase its efficiency. In many of these efforts, government funding covers the bulk of the work, but Don typically has to pony up a sizable cost-share, and fit the numerous meetings and extra work in and around his usual farming tasks.

With all of the agency people and researchers that he has partnered with, Don is not afraid to speak his mind. "I don't agree with everything," he says. "You have research that comes in and there are some things that aren't necessarily so. It's hard to contradict a researcher, but sometimes you have to. When it's not so, it's not so." He continues, "I feel I have lived here long enough, I should know a little bit about the stream itself. You have individuals who have had schooling on these resources, and that all helps. When you put us together, you've got a collaborative program."

INNOVATIONS

Successful new ideas emerge when people with a passion for their work encounter a significant challenge and begin to dream a bit.

Some of the people featured in this chapter stumbled by chance onto their path, serendipitously hearing the right thing at the right time, or discovering something unexpected. For others, a lifetime of experience led them to question accepted norms. In all cases, at some point, a leap of faith was required, a belief in what could be.

The people in these stories are developing and testing ideas that stand to improve the ways we build houses, produce food, and restore and manage landscapes in Washington and beyond. The fruits of their entrepreneurial spirit and savvy will provide benefits today and into the future.

FOREST CONCEPTS, LLC

Jim Dooley acknowledges that Forest Concepts' product line, despite its careful engineering and serious purpose, has a rather playful association for most newcomers. "Everybody that comes in here calls it Tinker Toys for adults," he laughs. But Forest Concepts' toys are larger scale than Tinker Toys, made from the long skinny trees that now dominate western forests. When put together, they make structures used for watershed and habitat enhancement.

Forest Concepts promotes forest health by using the byproducts of forest restoration in constructing their products. Thick stands of small diameter trees pose a significant fire danger for many western forests, and thinning out the trees is an oft-prescribed but costly solution. Forest Concepts is working to increase the markets for small timber, making needed forest restoration more financially viable.

Forest Concepts develops products with an eye toward completing what Jim calls "the watershed cycle." He explains, "We capture small diameter wood and other biomass from the forests and watersheds of a community, use minimal processing, and turn it back into functional products that can be used in that same watershed. We look at the materials imported into watersheds and replace them with local materials, whether it is picnic shelters, fencing, habitat, or erosion control devices."

For example, salmon habitat restoration plans often call for a large stump or log to be added to stream sites. Particularly in urbanized areas, finding and installing an old growth stump can be quite a daunting and expensive prospect. Forest Concepts has developed an alternative: ELWd® (pronounced "el-wood"), an assemblage of small logs fitted together to create something that looks and acts like an old growth log in a stream.

Like Tinker Toys, ELWd® logs rely on sticks fitting snugly in predrilled holes. With a bit of coaching, the relatively easy on-site assembly can be completed by untrained citizen volunteers, with minimal impact on the site. Only wood is used to build the structures, so when they eventually decay, no metal is left behind in the environment. Because the ELWd® structure has a lot of gaps and crevices that collect organic material and seeds, they may actually be more effective for restoration efforts than the solid logs they are designed to replace.

Close ties with research universities and government laboratories have contributed to the development of Forest Concepts' innovative products. One current federal research contract is for development of WoodStraw™, an erosion control material for disturbed ground intended to replace imported straw that can introduce noxious weeds and pesticides into the environment. "We spent two years figuring out the science behind why straw does what it does and then created a wood-based material," explains Jim. "Forest Concepts' version of straw is ecologically preferable and longer-lasting than agricultural straw."

Another government contract supported the development of wildlife-friendly fencing made from small diameter logs. The fencing Forest Concepts developed provides an alternative to barbed wire. It uses local timber and doesn't require posthole digging or metal fasteners. In addition, Forest Concepts integrated wildlife science in the design. Jim explains, "The spacing and locations of the rails allow antelope and juvenile wildlife to go under, and the top rail is low enough that deer can go over."

RENEWING THE COUNTRYSIDE

Forest Concepts has ELWd® models that replicate log jams, nurse logs, floating habitat rafts, and "critter condos." And then there is a line of garden products like rustic roundwood planters and small logs for creating terracing. More retail products are in the works, such as ready-to-assemble round wood structures—gazebos and hot tub and picnic shelters—for the do-it-yourself customer. Eventually, an in-store computer system will allow customers to design and purchase what they need at their local home improvement center. "Say you want a boat shelter, you can lay it out and the computer will put together the parts list for you and the drawings on how to put it together," Jim says. "You put it all in your pickup truck and go home."

Forest Concepts was founded in 1998 by Jim and several colleagues who had spent most of their careers working for large timber companies. "We saw the limitations of the corporate industrial model where you have this big complex, and either you operate it or you close it," explains Jim. "The timber towns really suffer when the big complex can't operate. There are still people in the town who want to be employed in forest products, but it doesn't make sense to have 500 employees inside one fence."

Changes in the forest ecosystems also were a factor in starting the business. "This company was formed to capitalize on the shift from old growth timber to small diameter wood," notes Jim. Forest Concepts has a clear advantage with public agency clients investing in restoration because their products are sold in local, domestic markets. The big companies are not positioned to be players on public land because they export logs and federal regulations currently prevent the export of timber that has been harvested from public land.

Forest Concepts relies on a decentralized network of licensed rural-based producers. "We decided to start with an outsource model where components are built by specialists," Jim explains. "You find smart people where they are and put them to work. We're bringing to rural forest industries a lot of the same business techniques used in microelectronics: pieces of the product come from all over, and our corporate entity manages the integration and customer relationships." Forest Concepts' rural producers are the ones who actually manufacture and sell the products, using the licensed trademarks and patented technology. The company is currently working with a handful of rural-based businesses and community groups in the Northwest, and is developing other relationships throughout the west.

The company is headquartered in an office complex in Federal Way, effectively preventing them from getting much into production. "Our location keeps us honest," says Jim. "We would have lots more fun on the production side, but we know where we can contribute most is in the development, marketing, and logistics." Thus far, Forest Concepts has reinvested its profits back into product development.

Forest Concepts also offers business development training, small wood utilization demonstrations, and specialized tools and equipment for handling, transporting, and processing small diameter logs. "Our technologies enable small entrepreneurs and rural communities to either create or expand a business around the use of small diameter timber for watershed uses," Jim remarks. "The complement of things that a community actually produces can be quite different because of the context of their community, watershed, ecosystem."

The bottom line, for Jim, comes down to some pretty simple questions: "Why truck steel fence posts and barbed wire in from 500 miles away when the community's got timber and people who want to work with it?"

GIES FARM

When Dale Gies decided 12 years ago to try rotations of wheat and mustard in between crops of potatoes on his family's farm south of Moses Lake, he didn't realize he was on his way to become a leader in the field of biofumigation.

Dale grew up on this 500-acre farm that his family has been intensively cultivating ever since irrigation allowed his father to break the ground out of sagebrush. Gies Farm primarily sells potatoes for processing, wheat and vegetable seeds, and grain corn. But over the course of 40 years, soil productivity declined due to wind erosion, low levels of organic material, and compaction. Dale decided to experiment with changing his cropping system to improve soil quality by increasing water penetration and retention and reducing soil erosion.

Mustard plants had been shown in laboratories to have fumigant qualities, but those findings had not been successfully replicated in the field. Dale comments, "Some of the green manure crops cause disease, host nematodes, or become a weed themselves. We knew mustard had the potential to help with disease and nematode control, but we didn't realize it could be as effective as it has been. Really, we just wanted to protect the tilth of our soil while we were growing rather intensive crops like potatoes and onions that don't produce a lot of residue."

Dale describes his first trials: "I let the mustard grow up four to six feet tall, then in October I would chop it up and incorporate it back into the soil before planting potatoes. When I went back in and compared where we chemically fumigated with where we used a mustard crop, I didn't see any difference." After three years of the same results, Dale contacted Washington State University, who sent researchers out to do replicated trials in his fields. He recalls, "As the results began coming in, the researches said, 'Wow, this mustard works as well as fumigant!'"

Dale has seen a 30-plus percent increase in moisture holding capacity in his mustard-managed soils compared to the same soils under conventional management. He reports, "The green manure does some rather unique things to the soil as far as structure and tilth, and with the chemicals in the mustard, we actually solved a number of weed, disease, and nematode problems." With his rotation of potato, wheat, mustard, back to potatoes—minimal tillage, stubble mulch, and green manure—Dale has seen organic matter levels increasing, something which is thought to be impossible over a potato rotation. "The soil just keeps getting better!" he says.

INNOVATIONS

149

Dale vividly remembers when researchers brought in an exotic wind machine to replicate and test the impact of strong winds on his fields. He recalls, "It looked like a giant vacuum; it has a big glass chamber you put down over the soil, and you keep turning it up and up and up, and you actually observe when the soil particles start to detach and move." Dale says with a smile, "They actually maxed the machine out on one of our fields that had had three green manure crops in the previous six years, and they couldn't get the soil to move."

When Dale and the researchers from WSU started testing mustard varieties, they learned of a research institute in Italy that breeds mustards for this purpose. Most of the work in the United States had involved using mustards bred for edible purposes. Dale recalls, "We brought in some of the Italian-bred mustards to test, and when we began producing those varieties ourselves, there was a lot of interest from others." Dale now markets this seed through seed dealers across the United States and Europe. He comments, "What started out as just something to keep our ground in good shape has become almost a full-time endeavor for me. I spend a lot of my time now working with researchers and dealers and growers trying to figure out how to tweak this technique to make it work in other areas."

Dale has reduced but not eliminated his use of fertilizers and herbicides. He does find that mustard helps to keep his nitrogen inputs low: "The mustard ties nitrogen up in the fall and keeps it in an immobile form all winter. Then in the spring when the soils warm up and you plant your crop, the nitrogen becomes gradually available to the plant, along with other nutrients." Dale carefully monitors water evaporation on his farm, and with judicious irrigation has almost eliminated movement of water below the root zone, reducing fertilizer waste and water pollution.

Fumigants have been eliminated entirely on the farm. Dale remarks, "That got people's attention. Not only could we grow a high quality product, but we are considerably over the county's average yields." He continues, "We don't have some foundation funding our work, this all has to fund itself. We found it just makes good economic sense. We are able to grow higher value crops, and we are doing it for less money."

Dale has invested some of what he has saved into creating wildlife habitat on the odd corners beyond the reach of the center pivot irrigation system; pheasants are his primary customers. He explains, "We tried to put together all the critical elements of wildlife habitat: native grasses for them to nest in; shrubs that provide winter cover and protection from predators; and some berries. We plant food crops—safflower and corn—so that when we get a really nasty winter, we don't lose all of our birds."

While Dale is enthused by what he has discovered on his farm, he doesn't expect that the industry standards will change overnight. "One of the things that researchers are finding is that it's hard to get people to do green manures. If you can tell them it will reduce disease, weeds, or nematode, they'll do that, but not just to improve the tilth of their soil." It pains Dale to see farmers baling and selling straw that won't bring in $20 an acre, while their soil loses $60 to $70 worth of nutrients on that same acre. "You've got to feed that soil and people don't realize that," he says. "There is more life in the 18 inches below the soil surface than in three feet above it. It's like Will Rogers said, it's not that we don't know anything, it's just that half of what we know isn't true."

Dale knows that farms like his are in the minority. He says, "In our opinion, if we have to destroy the ground to make a living, we probably ought to go look for another job."

IRONSTRAW GROUP

"Homes Cows Would Love to Eat" was the title of the newspaper article that changed the lives of Michael and Spring Thomas. They were both teachers at a community college in Seattle when they came across a story about the revival of the age-old practice of strawbale construction in New Mexico. While they were intrigued by what they read, they had no way of knowing it would lead them to bring strawbale building into the mainstream in Washington state and beyond.

Straw is a plentiful, renewable resource that makes an excellent, affordable building material. Walls are made of two-foot wide, 80-pound bales of straw that, once they are stuccoed, are more fire-resistant than ordinary wood-framed walls. Strawbale structures are rugged, and can withstand winds of 100 miles per hour. Houses made of straw are well-insulated, and they last. The first strawbale homes built in Nebraska in the late 1800s are still in use.

Michael and Spring began to dream of building their own house of straw. Michael remembers, "All the information on strawbale building at the time, from 1991 through 1993, was that it was being done in the drier Southwest. We thought we might move down there and build our house. We were doing research and

investigating this for a couple of years before we finally heard that, in fact, this was being done in Norway. And we looked at each other and said, 'Well, gosh, if it's being done in Norway, why can't we do it here?'"

They soon found that the obstacle to building strawbale houses in the Northwest had nothing to do with the weather. Local building officials on the Olympic Peninsula where they had relocated were reluctant to approve the unfamiliar building technology. Michael remembers, "They had no idea about strawbale buildings at all. They just thought it was the three little pigs' house. Our plans were fully stamped and approved and architecturally designed— which is normally satisfactory for building officials—but they just didn't know what to do with them."

After five months of hearing "no," they compiled their research on the economic and environmental advantages of strawbale construction, and founded the IronStraw Group to disseminate what they had learned. Says Spring, "As educators, we shared with people because that's who we are." They also hoped that having official letterhead and brochures would with help with their legitimacy in the permitting process.

Spring remembers in 1994 when they founded the non-profit, "We had never been in a strawbale building. There were no books. There were no videos. We just believed in it. It made so much sense." Michael adds, "It was a leap of faith. Since we were educators and advocates, we didn't want to use the experimental provision of the building codes, or do anything other than get a fully approved, permitted building." In the fall of that year, Spring and Michael finally got their plans approved. More than 60 people came together to raise the walls of the first fully permitted, load bearing strawbale house in Washington State. But Spring and Michael were just beginning their lifework to make strawbale building a credible and more widely-used method of construction.

During its first four years, IronStraw helped in the construction of more than 25 strawbale houses in western Washington before Spring and Michael decided to move closer to the source of straw. Soon IronStraw had forged partnerships with over 50 community organizations and corporations in the eastern part of the state, and laid plans for a second demonstration strawbale house in Cashmere. This time they incorporated another goal into the project: teaching building skills to at-risk, disadvantaged rural youth. Says Spring, "These kids were a large part of the joy and trials of working on that project. But to hear an 18-year-old who initially wouldn't answer a direct question burst out in song as he's pounding a hammer makes all the frustration melt away."

IronStraw has continued to integrate youth empowerment into their projects, working with social service organizations and the Yakama Nation. Michael explains, "We actively pursue these partnerships and look for how we can have youth involved in all of our projects. Our focus is community-based building. We work

to empower people from community organizations during our wall raising weekends." Some of the young people that have developed skills working on IronStraw projects have been able to leverage that experience into jobs with contractors.

The affordability of strawbale building has always been one of its stronger selling points. Michael estimates that an owner willing to put in some sweat equity, and store and reuse materials, can build for $40 per square foot, which is comparable to the cost of Habitat for Humanity housing. A conventional house built by a contractor runs more in the neighborhood of $80 per square foot.

In addition, getting rid of straw is a big challenge for wheat farmers in the region. After harvesting the grains, Washington State farmers are left with an estimated 7.5 million tons of straw in their fields, and because of the strength of the fibers, it does not readily decompose. In 1999, farmers were fined tens of thousands of dollars by the state for illegally burning straw. Thirty members of the Association of Wheat Growers toured IronStraw's demonstration strawbale home in Cashmere, attracted by the prospect of selling

building-quality bales of straw for two dollars a bale.

The Wheat Growers teamed up with IronStraw and the Washington State Office of Community Development to build housing for migrant farm laborers. A 1999 report from the Washington Governor's office found that about 60 percent of the migrant workforce in the state lacks housing during the growing season. After a fortuitous meeting with a willing orchardist, IronStraw set to work alleviating that problem. They built three new strawbale buildings among pear and apple orchards outside of Omak.

More than 100 local youth helped raise the walls of the homes in just three weekends. Each of the six units has multiple, good-sized bedrooms and bathrooms. The thick muted copper-tinted walls and shiny new appliances compare favorably to most college dormitories, not to mention the thin shacks hidden behind orchards elsewhere that pass for housing. While the tenants will be transitory, the buildings seem likely to outlast most of the construction in the rest of the neighborhood.

Michael and Spring don't just sling bales of straw around. "We have talked with all 39 county building officials in the state of Washington," explains Michael, "and now every single county in the state will approve strawbale building. The building officials know we can provide plan review, and structural and construction details to make sure it's done correctly." Over the past five years, IronStraw has lead over 90 workshops, seminars, classes, and presentations, reaching people from as far away as Ireland, Japan, and Australia. Habitat for Humanity has taken advantage of IronStraw's educational programs, and seven chapters now use strawbales in constructing affordable housing.

According to Spring and Michael, strawbale construction is a very simple technique to learn. To date they have been involved in building upwards of 55 houses— plus a 5000 square-foot school, a greenhouse, and a library. Says Spring, "During our community-based building workshops, 25 unskilled people who don't know anything about strawbale building can put up a 1200 square-foot home in a weekend. People love building together. When people see how easy it is, they say, 'I can do this, I can build my own house!'"

RENEWING THE COUNTRYSIDE

LOPEZ COMMUNITY LAND TRUST'S
MOBILE MEAT PROCESSING UNIT

Pastoral Lopez Island seems an unlikely site for a revolutionary innovation in small farming. Northwest Washington's San Juan Islands are best known as a bucolic weekend retreat for frazzled urbanites. But with the help of a local community land trust working well outside of the box, island livestock growers now have access to a USDA-approved mobile meat processing unit—the first of its kind in the US—that is attracting the attention of agriculturalists across the globe.

With its tourist-based economy, vacation homes, and limited land base, San Juan County suffers from the largest income-housing "affordability gap" in Washington State. Over the past 15 years, the Lopez Community Land Trust has built three clusters of modest but appealing single-family homes for lower-income island residents. According to former executive director Sandy Wood, "This community land trust is one of about three in the nation that have a broader mission, so we also work in the areas of sustainable agriculture and rural economic development."

"Housing is connected to livelihood and making a livelihood on the land is a major issue in this community," says Sandy. "Affordable housing was an insufficient response to a larger systemic problem.

So interest grew in supporting small-scale agriculture as one of the most desirable and viable year-round economic bases for the islands." In 1996, when island farmers and the local Washington State University extension agent began talking about increasing opportunities for local food processing, the Lopez Community Land Trust saw an opportunity to implement the parts of its mission that extended beyond affordable housing and the confines of one island.

As recently as 2001, it was illegal for a sheep farmer on Lopez Island to sell his neighbor a lamb chop. A farmer could only sell meat directly by the whole, half, or quarter animal, requiring custom slaughter and a lot of freezer space. Most individuals, restaurants, and retailers are only interested in buying meat by the cut. For farmers on the San Juan Islands, the closest USDA-approved slaughter facility was 200 miles away by land and sea. Most island livestock producers would just sell their animals at auctions on the mainland for a fraction of their retail price.

The mobile meat processing unit has provided an efficient alternative that increases the return and local customer base for farmers, and allows consumers to support small island growers. From the outside, the 26-foot mobile meat slaughter unit looks like the long white trailers hauled on 18 wheels that are ubiquitous on our freeways. But this particular trailer is attended by a USDA inspector and is equipped for the on-farm slaughter of cows, sheep, hogs, and goats. The Land Trust has retained ownership of the facility and leases it for a nominal fee to the Island Grown Farmers Cooperative, which was formed for this purpose.

Bruce Dunlop, an apple and sheep farmer from Lopez with an engineering background, has been intimately involved in the development of the mobile processing unit, as a community volunteer and a consultant. He explains, "There was really a three-legged partnership here in the county between the Community Land Trust, the extension office, and the local producers that formed a steering committee."

A feasibility analysis was conducted with producers and consumers. Bruce reports, "The key findings from the consumer studies were that consumers definitely wanted to purchase local meat products; they wanted to be able to buy small quantities, and they wanted convenient access, which really means farmers' markets and grocery stores." With this information in hand, they forged ahead. Bruce laughs now, "We didn't know any better."

Says Sandy, "There was absolutely no guarantee this thing was going to get the USDA grant of inspection. It was a big gamble. The program branches of the USDA were very supportive and provided much of the funding for it. But the regulatory arm of the USDA was not at all enamored with the project." The USDA's concern was legitimate because the implication for them was, if the facility was approved, they would have to provide an inspector. Sandy reports, "But we met their specifications, and they awarded us the grant of inspection. Now I think everybody in the country wants one!"

According to Sandy the trailer and the tow vehicle cost just under $100,000. From start to finish, Bruce estimates the project cost about $350,000, though that doesn't include countless hours of volunteer time. During the extensive testing period, the mobile processing unit lived in Bruce's driveway. There were a few glitches

to work out. In fact, they had to replace their whole hoist system right away because it was inadequate for large steers.

Of course slaughtering an animal doesn't quite make it consumer-ready. Bruce explains, "When the mobile unit leaves the farm the carcasses are all hanging in a cooler being chilled as whole carcasses or quarters. Beef needs to age for two weeks, so it goes someplace where it can sit in a cooler. Then it has to be cut and packaged into consumer-friendly sizes, and that needs to be done under USDA inspection as well. We basically realized that we needed to also operate that cut-and-wrap facility."

After weathering considerable local opposition in trying to identify an appropriate location on Lopez to build such a facility, the Farmers Cooperative found an easier option not far from the mainland ferry terminal. "We located a suitable facility that wasn't being used that we were able to lease," says Bruce. Since this USDA-approved facility was outside San Juan County, the Island Grown Farmers Cooperative expanded its membership to 40 to include farmers in Skagit, Whatcom, and Island Counties.

Sandy says, "One of the things that surprised me was when we first started up, we were getting so many calls—way before the unit was ready to go—from high end restaurants in the Puget Sound area. 'When is the meat going to be ready?'" While those calls still come in, the vast majority of the island meat has stayed on the islands. Some Island Grown Farmers Cooperative members on the mainland sell their meat in local grocery stores and through Seattle-area farmers' markets, and the cut and wrap facility in Bow offers retail sales two days a week.

The slaughter truck and the cut and wrap facility employ six people full-time. The facilities are financially sustained by membership and user fees. Says Bruce, "We are currently processing 40 head of beef a month and our aging and cutting facility is at capacity."

The mobile meat processing facility has also had an impact on how farmers manage their land and animals. Sandy explains, "If you're going to sell at the auction on the hoof, you have to have a huge number of animals to have any kind of income at all, which is hard on the land. Now that farmers can carry their animals all the way to slaughter-ready, they can have fewer cows, rotationally graze them, and grass finish them rather than sending them to the feedlot. Because of the processing unit, farmers are able to change their practices in ways that not only support a healthier physical environment, but also give them larger profits."

For the consumers, there are unexpected reasons to buy Island grown meat. Animal cruelty prevention groups are interested in the mobile unit because it minimizes the animals' suffering. Explains Sandy, "The unit and the animals are right there in the pasture. They're stunned in the field where they're sitting happy as a clam—or a cow! And they don't have the trauma of transport or the slaughterhouse environment. From a meat perspective, when animals are traumatized before slaughter, their endocrine system responds in such a way that it toughens the meat."

And if that's not convincing enough, Sandy says the USDA inspectors who have worked in the mobile unit are impressed: "They say this is the healthiest meat they've ever seen. They don't really enjoy working in the packing houses. So to be out in the field and see really healthy meat and healthy animals, they love it. On a personal level they're very supportive."

Sandy reflects, "We didn't realize what a big splash this was going to make. We didn't know that the problems of small farmers across the nation were so much like ours. But what we've learned is that the infrastructure for small-scale farming is just evaporating everywhere. The smaller slaughterhouses in rural areas are closing down; they're disappearing. So the small producers are left with fewer and fewer options."

The good news has spread quickly and groups around the globe have inquired about the project. Bruce is now working part-time as a consultant helping others launch their own slaughter facilities. "I think we will see more small-scale slaughter units coming on line" he says. "Farmers are looking at how to control their own destiny. They want to sell direct to the end consumer as opposed to on the commodity markets where you have to be big to succeed."

PESHASTIN CREEK
GROWERS ASSOCIATION

Central Washington's Wenatchee Valley is an excellent place to grow pears and apples, with warm dry days and cool nights. Around Peshastin Creek, a tributary of the Wenatchee River, a handful of family orchardists are carrying on the work started by their grandparents and great grandparents. But this generation of pear growers is experimenting with gentler pest management strategies, and it seems to be working.

Dr. John Dunley is a Washington State University tree fruit entomologist, and the scientific names of bugs and words like "organophosphates" roll smoothly and quickly off his tongue. His work has involved ploys that foil the reproduction of one of the key pear and apple pests: codling moth, which he describes as "the proverbial worm in the apple." By bathing fruit trees in female codling moth sex pheromones, somehow—and no one really knows how—codling moth mating is delayed and the result is fewer fertile eggs. This happy discovery means that there is less need for chemical insecticides.

This codling moth mating disruption program has now been effectively applied on half of the acreage in Washington. In 2000, John set out to build on this integrated pest management success story. He wanted to complement the mating disruption approach with strategies to increase the populations of beneficial insects that eat codling moths and other pests.

John explains, "A lot of the natural enemies—or good bugs—come from the surrounding natural vegetation, so if everyone went to a

soft (low pesticide) program, we could create avenues for movement of natural enemies over an entire area without any pesticide barriers." Many pesticides are indiscriminant and kill both the bad and the good bugs. John was still developing his idea when the Peshastin Creek Growers Association got wind of it. Says John, "I was going to take another year to further develop the theory and the economics of it, when these guys heard about it and gave me a call."

Dennis Nicholson would be a contender in a Hemingway look-a-like contest. He is a third generation orchardist and a member of the Peshastin Growers Association that is made up of ten third and fourth generation orchard families. Dennis explains, "There are a couple things going on here that made us willing to try this. One thing is the creek that flows through the area. We know that we don't want a lot of pesticides dumped in the creek. Worker safety is another thing we are concerned about. Finally, we have high visibility because of the highway; a lot of people come through here and they want to know what we are spraying." A third of the acreage had previously been certified organic.

Even before they met John, the Peshastin Creek Growers Association had been proactive, implementing water conservation measures that allowed them to give back a quarter of the water rights they were originally allocated, increasing flows in Peshastin Creek. John and the Association began meeting weekly to discuss his ideas. He recalls, "You don't really know how much it's going to cost, if it's possible, if you're going to cause more problems than you're going to solve. We weighed the potential benefits versus the risks. Eventually we came to a consensus: everyone agreed that we'd give it a shot."

In the beginning there was a lot of concern about whether or not there would be economic damage in the orchards due to pests coming in and harming the crops so they couldn't be sold. Dennis says, "In some situations that did happen because people were not on top of things. That has been part of the learning curve: you have to farm a little differently in a soft pesticide program than you do in a conventional program."

All the orchardists were surprised by the additional time required. Soft insecticides actually require more frequent applications because they target specific life stages of specific insects, rather than killing all bugs all at once. "It was a steep learning curve for me," Dennis admits. "I was used to putting on one spray and walking away for about 30 days and not having to worry about spraying a second time. It turns out we are probably spraying about four times in that 30 days with the softer and organic programs. So it's much more management intensive. It took me a while to get up to speed and to start looking for the signs of when we needed to spray and having myself prepared and ready to go."

The extra work is paying off, as other pear farmers are finding that broad spectrum pesticides are no longer effective with some pests. John says, "If you go over to our conventional orchards we use for comparison, in some of those orchards we have put on a large number of broad spectrum insecticides, and they still are suffering

significant economic damage. These guys are sitting pretty compared to the rest of the valley."

Now half of the 310 acres managed by the Peshastin Creek Growers is organic and the rest is managed with a soft program. Says Dennis, "It's gotten to the point where the soft program is very effective. The spray I am currently using for the soft block is exactly the same spray as I am using for the organic blocks. The only significant difference between the organic and the soft block is that we are able to use herbicides, rodenticides, and fertilizers."

The Peshastin Creek Growers collectively market under the "Gently Grown" label and received the Food Alliance sustainable agriculture seal of approval in 2002. Finding markets for their unique product that is not quite organic and not conventional has been a bit of a challenge. Says Dennis, "Last year for the first time we found a customer in California who was willing to pay us a premium for the Gently Grown label, so there are niches out there that are willing to pay for a pesticide-free or low pesticide fruit."

"One of the things that might be different about our group is that we all started out as conventional farmers, and our fathers were conventional farmers," adds Dennis. "We had to come into the idea of a soft pesticide or organic program from our fear of insect resistance, our fear of polluting water, our fear of regulation, plus our motivation to raise a superior product and do it in a kinder, gentler way. That's where the Gently Grown label comes from; we are trying a more 'gentle with nature' approach to growing our product, and basing our inputs on need."

QUILLISASCUT FARM SCHOOL OF THE DOMESTIC ARTS

LoraLea and Rick Misterly have devoted their lives to preserving small-scale farming traditions and providing discriminating customers with unique, high-quality products. With Quillisascut Farm School, the Misterlys have gone one step further, providing hands-on education for aspiring chefs—education that demonstrates the value of locally-sourced, small-scale food production.

The curriculum at this pastoral school includes goat milking, cheese making, chicken plucking and eviscerating, wild edible plant gathering, and lamb sausage making. The urban chefs and culinary students who gather here may know a lot about the later stages of food preparation, but at Farm School retreats they experience—first hand—the work of cultivation that is usually taken for granted.

Quillisascut Farm is named for the creek that winds through scenic Pleasant Valley before it drops into the Columbia River in the northeastern corner of Washington. The valley used to be called "Peaceful" before warring neighbors necessitated the switch to "Pleasant." These days the wild turkeys are the only ones disturbing the peace in the valley, at least until the goat kids arrive in the spring.

Fifty goats and a Jersey cow range over much of the 36-acre farm. LoraLea and Rick strive for self-sufficiency and grow most of their own food. In addition to the dairy, LoraLea and Rick's farm has a small vineyard, fruit trees, two commercial kitchens, organic vegetable and herb gardens, a wood-fired brick oven, and an energy-efficient strawbale bunkhouse and dining room.

LoraLea has been selling her handmade goat cheeses to top Seattle restaurants for over a decade. She first invited chefs out to Quillisascut Farm as a way of thanking them. She explains, "I've really been appreciative of the chefs that use our products, that they are willing to take the extra time to buy from a lot of small producers, instead of just going to some big wholesaler. So I thought, how can I give something back to that community?" A farm visit/retreat seemed just the thing, and in 1993 the first "fude ranch" brought Seattle chefs out to the foothills of northeast Washington's Huckleberry Mountains for farming, wining, and dining.

The success of that first retreat led the Misterlys to expand the occasional farm retreat for restaurant staff to a more structured on-farm school for those in the culinary trade. Rick explains, "Our first idea was to get the chefs out here, have them cooking with fresh ingredients, and get all excited about it. They were excited about it, but as they got more famous, it was harder to get them out here." Culinary students have proven to be the most receptive audience, and several culinary schools now offer scholarships. The Quillisascut Farm School of the Domestic Arts "introduces culinary students to the source of their work: from the farm to the table." Small groups spend seven days on the farm in the summer, learning by doing farm chores and preparing elaborate meals with ingredients they have harvested.

"Farm Culinary 101" focuses on the early stages of food preparation, what happens before cheese, meat, and produce arrive in a kitchen. Farm days start at 5:00 AM in the milking shed. Presentations on heirloom plants, seed saving, and the benefits of grass-fed meats are interspersed with wild food walks and garden glove tasks like composting, transplanting seedlings, and building raised beds. Farm students visit other local growers to learn about beekeeping and orchards.

to find a source of meat that is grown in a respectful way."

Every afternoon and evening, students work to outdo each other preparing culinary delights using the fruits of their labor: sweet corn

Slaughtering animals may be one of the more daunting aspects of the farm school agenda, and also perhaps one of the most transforming. "People have this horrible mentality of blood and gore, but that's not what it is about. The inside of the animal is all very orderly and it's a respectful process," says LoraLea. "A lot of the students leave feeling better about eating meat, and they want

flan, lamb terrine with chokecherry mustard, purple potato gnocchi, lavender honey ice cream. Says Rick, "They see what it takes for a small farmer to get products from the seed or the tree to the market. I want them to realize why a farmer might charge more than a big distributor. But what you're getting for that extra cost is the freshness and the difference in flavor."

LoraLea grew up eating cheese her mother made by hand in

Leavenworth, Washington: "I remember the taste of fresh curds, real creamed cottage cheese, and butter. It is a taste that isn't duplicated in anything found at the local grocer." When she and Rick first moved to their land in 1981, they lived in a tent with no

have been making cheese for centuries," LoraLea says. She strives to teach American culinary students an approach to food that is more common in small European communities, where production speed is not valued over quality and tradition, where seasonal, locally grown ingredients are the centerpiece.

The Misterlys offer retreats tailored for students, professionals, and food lovers of any persuasion. Culinary students aren't the only ones who are disconnected from their food source. Local kindergarten and elementary school kids have been visiting Quillisascut Farm in the springtime for over a decade. Rick explains, "When they first started we thought, 'Well, that's kind of stupid. These kids live in this area, they must know all of this stuff.' And then you realize that they don't." In fact the Misterlys are one of the few

electricity while they built their house and outbuildings. LoraLea learned to make cheese in their outdoor kitchen, storing it deep in their well to keep it cool in the summer. She now has a state-approved kitchen for cheesemaking, and produces about 5000 pounds of cheese a year.

families in their agrarian community that rely entirely on the farm for their income.

The Misterlys style harkens back to an earlier time, and perhaps signals a wave of the future. "I am interested in the way people

The Misterlys enjoy sharing the bounty of their farm and their passion for sustainable agriculture, and having extra farm hands and income has helped to keep their operation viable. Says LoraLea, "There are a lot of things that are missing when you just go to the produce section of the grocery store. It's about passing on our tradition."

Rendezvous Reclamation

A new breed of settlers have been attracted to the Methow Valley in recent years, looking for their piece of rural paradise. They usually don't bring much expertise in land management, but they have a sincere desire to preserve the agricultural look and heritage of the place. Often the farmland they buy has been left fallow for years, leaving plenty of room for noxious weeds to take up residence. Sam Lucy has built a business to address this problem, reclaiming abandoned farmland using crops rather than chemicals to bring weeds under control and the soil back to fertility.

Sam moved to the Methow Valley in 1992, but unlike many newer residents, he brought a lifetime of farming experience with him. The seeds of Rendezvous Reclamation were sowed six years ago, when Sam was doing custom planting work for a local rancher with a weed problem. "The regular way to deal with knapweed is to spray," he explains, "but I thought there might be another way." Time and again he had seen healthy crops crowd out weeds—particularly grains. Using his employer's equipment, he experimented on his own 20 acres that were infested with diffuse knapweed. He says, "I disced up the ground, seeded some spring rye, then stood back and watched my knapweed field become a beautiful field of rye. Boy wasn't that neat!" As it turned out, spring rye had been grown all over the valley for hay for many years.

While Sam was conducting experiments on his own ground, he continued as a hired hand elsewhere in the valley. "Various folks would stop by while we were plowing other fields in the spring and ask if we could come do their field," he recalls. Sam knew that he couldn't own a home, raise a family, and eat year-round in the Methow Valley on seasonal farm-hand wages, so in 1998 he decided to launch his own business, named Rendezvous Reclamation. "It took a leap of faith," he admits. "I'm quite sure I wouldn't have the guts to do it again, but here I am, and most of my equipment is nearly paid off."

"Like many mountain valleys in the West, abandoned farm fields are as common as March snow squalls," explains Sam. What is happening here, however, is that second home owners—mostly from Seattle—are buying up these lands and building their vacation homes. They are then left with the need to care for their "open space." Sam notes, "Many of these folks actually have a good land ethic, and want to see their weed patches brought back to health. And many of these folks are wise to the dangers of chemicals, despite constant reassurance from the local county weed board."

Rendezvous Reclamation is Sam's alternative to the 50-year chemical war on weeds that he has seen promoted by federal, state, and county agencies here in the West. He explains, "As I became more aware of all the chemicals used in conventional food systems throughout the Columbia Basin and all across the wheat country, I was alarmed. Then when I saw the onslaught of chemicals used by the Forest Service, State Department of Fish and Wildlife, the State Department of Transportation, and certainly the county, I was flabbergasted! I thought aerial broadcast spraying was a thing of the past. That's how naive I was! I knew there had to be a better, longer lasting, more ecologically sound way to promote healthy landscapes."

Thus far Rendezvous Reclamation has served 40 clients, managing land that ranges from two to seventy acres. Many of Sam's clients want their ground returned to a so-called "native state." A typical restoration will involve Sam discing or, in some cases, plowing up the land. Then he seeds a crop of straight aggressive grain, either spring rye or triticale. Sam explains, "This gets a jump on the knocked-back weeds, has good biomass volume, and is fairly drought-resistant. The first year is like magic, but then the work begins."

In the second year the cover grain is turned under and Sam evaluates the weed situation in the fall. Then he plants a blend of native bunchgrasses and fescues on a roughed up seedbed, and hopes for next year's rain. He reports, "On ground that is not irrigated, rain is the obvious wild card—the one that makes all the difference. I had the luxury of starting all this at the beginning of a four year drought. This year we got some rain, and plantings I did even four years ago suddenly began showing strength."

Because grasses take at least a couple years to establish, Sam keeps volunteer grain and weeds clipped off with a mower, which keeps them from going to seed until the grasses have taken over. For many weeds, there are now proven bio-controls available, such as knapweed weevils that he sometimes uses in place of mowing.

Sam grows crops for harvest on some of the land that he manages. "Each year I try to pick up another field for actual crop farming," he explains. "I grow a rotation of certified organic grains and flax alternating with green manures of clover and peas on the irrigated farm ground." Sam and his wife Brooke are hoping to direct market some of the crops he grows soon.

"Lately I've been growing two very old grains on contract," Sam explains. "Emmer is a spring-seeded grain that is also known as farro; it is primarily grown in Italy at this point. Spelt is a fall-seeded grain. Both are thousands of years old and are favored by folks who have developed allergies to the modern wheats, which have been bred to the point where they are too complicated to digest." These grains are not only beautiful in the field, revenue for emmer is twice that of wheat, in part because it is currently grown only two other places in the United States.

"At heart, I am a farmer," says Sam. "Our services offers several advantages to landowners: they get to keep their water rights; they get to keep their land under agricultural use status which is a significant tax break; they get their land well cared for; it looks beautiful; and they are responsible for keeping local agriculture alive. Agriculture is a vital part of the beauty of the valley for all, and the idea of getting back to local foods is becoming an increasingly palatable idea."

Sam believes the very best scenario is one in which viable farms once again become part of the Methow. He suggests, "Perhaps more young farmers will be attracted to the valley if a market is proven for creative crops. It isn't necessarily re-inventing the wheel, but

more getting back to the old wheel. Conservation easements are a beginning, but by themselves, they certainly cannot maintain a vibrant landscape. I like to think that family farms can remain the backbone of a rural landscape."

"My main motivation remains the land—my love for the land," says Sam. "I hate the thought of tireless spraying programs that have no long-term gains at all, particularly during the spring when it's nesting season. I love the thought of a land full of creepy crawlies, bird song, poignant smells, and a view that brings a smile. There is no thrill in herbicides and deadened landscapes. I'm very, very fortunate to be able to work with and for many folks who feel the same. By promoting a system that works with, instead of one that battles nature, I truly hope to help keep the land healthy for all our children."

STILLAGUAMISH OLD CHANNEL RESTORATION

Sometimes complex problems have simple solutions. For 30 years proposals have been floated to improve salmon habitat in the Stillaguamish River's old channel, known as "Old Stilly." And for 30 years, every proposal was deemed too expensive or too impractical. Then local flood control commissioner, Chuck Hazleton, started experimenting with more efficient ways to control the flow of water, and the community came together to create a homegrown solution.

Chuck is a salty character with a bristly beard and a pipe often clamped between his lips. He worked building docks and tunnels before he became Flood Control District Commissioner for Snohomish County, which requires him to spend too much time in meetings for his taste. The old channel restoration project got him back to the construction projects he enjoys.

Until flooding altered the course of the Stillaguamish River, the old channel was the river's last eight miles to Puget Sound. Chuck explains, "In the old days, they took the channel and straightened it out so they could bring logs down it as there was considerable logging activity up on the hill. Why they chose this path is beyond me when there was a much shorter one." With flooding in 1912, the river took the more direct two-mile course to the bay, leaving the old channel to stagnate during the dry months, often stranding out-migrating juvenile salmon in what Chuck calls "a deoxygenated dead zone."

The dairy farmer with property at the mouth of Old Stilly suggested the installation of a water control structure to capture the tidal surge, reconnecting the old channel to the river system and providing much-needed side channel rearing habitat for endangered salmon. The project was funded by the Washington Salmon Recovery Funding Board and co-sponsored by the Stillaguamish Flood Control District, the Stillaguamish Tribe, and Snohomish County.

The Army Corps of Engineers was an early partner, but their plan turned out to be like others that had been proposed before: too complicated and too expensive. Chuck laughs as he recalls their estimate, "They got the price of it up to about $435,000, and we didn't have that much money so we fired them." A local engineering firm, Chinook Engineering, designed a smaller scale alternative—a concrete structure that stretches across Old Stilly with tidegates that close as the tide recedes, pushing water down the eight miles of old channel and eliminating the seasonal dead zone.

Getting permits was the next hurdle, and it proved to be sizable. "Project planning took three years and a lot of hydraulic studies," explains Chuck. "It amazes me how much money can be spent on engineering a pile of dirt." There was also some resistance from public agencies. Chuck turned his attention to the design of the tidegates themselves, and developed a plywood prototype that seemed a significant improvement over the much heavier tidegates typically used. "I started playing around with plywood and ended up with ultra high molecular weight polyethylene and high strength rust resistant alloy steel," he explains.

The construction phase of the restoration project was logistically challenging and relied heavily on local volunteer efforts. "We had

six days to build this thing between the tides," recalls Chuck. "The very first thing that happened was our coffer dam failed and we had to find different ways to do things. We called up the local quarry at seven o'clock in the morning saying we are in deep trouble, we need another 400 ton of big rock. Half an hour later the trucks were showing up. Nobody else would have done that, but they wanted this to succeed."

The project had the support of the entire community. Chuck shares one of many examples: "One of the guys that lives in the neighborhood came by and saw that we needed help. He called into work and said, 'I'm not coming in today,' then came down here, climbed on a piece of equipment and went to work. People weren't going to let it fail." Chuck put in a lot of volunteer time on the project himself. "Because I'm a commissioner with the Flood Control

District, I can only be paid to go to meetings, so they got a lot of free work out of me," he laughs. Other local volunteers worked replanting Old Stilly's banks and clearing invasive blackberries and reed canary grass to make way for shade trees.

Chuck is clearly proud of what was accomplished—neighbors working together and a final result that cost less than half of what the Army Corps proposed. "Now throughout the whole low tide cycle there is a flow of water where it used to be just sandbars and pools," he explains. The project also brought together constituents that didn't always see eye to eye.

Chuck's lighter tidegates were an important contribution to this breakthrough. "The most significant change is the way the hinge mechanism works," he says. "It's what allows the water to pass more freely." Chuck took a fish scale and pulled on the gates when he had them hanging in the shop, and found that it only took seven pounds of pressure to open a gate a foot. He explains, "With the old gates—most are cast iron—you had to have a tremendous amount of pressure on them to open, so when the water did hit, it spread out wide and was diverted out." The new tidegates help keep the channel clean because there is a constant, direct stream—benefiting farmers and the fish. "It is just incredible how changing a hinge mechanism can make a difference!" remarks Chuck.

Once the gates on the old channel project were replaced, it was clear there were other applications for the tidegates. Chuck explains, "We had the material, the district was going to replace some other gates, so the Tribe and community joined together again." The second project involved replacing the tidegates in a nearby drainage structure designed to prevent saltwater from intruding into farm

fields, while still allowing for fish passage to over 14 miles of upstream habitat. Chuck reports, "We are getting the chum salmon back. They have been planting chum upstream for years and not getting anything back, and all of sudden we change the gates and get a few back! The new tidegates give farmers what they need in drainage, and give the tribes unobstructed access for their fish."

The Stillaguamish Flood Control District has constructed nine of the new improved tidegates so far. "I kid about it being patent pending," Chuck laughs, "but I don't think, practically, you could do such a thing. I figure if I want to build these things, I'll just build them cheaper than anyone else who wants to fool with them." The potential for the tidegates is broad—applicable anywhere with tidal influence and drainage. Chuck envisions that tidegate technology will continue to improve now that it has been given a much-needed nudge.

AFTERWORD

When Sustainable Northwest first presented their project *Renewing the Countryside: Washington*, I was immediately drawn to the idea. Over the years I have learned the extraordinary power of good storytelling to motivate and energize people that can lead to a change in their behavior and attitudes. Reports and statistics are vital to help us understand the science behind solving natural resource issues, but are not complete without the compelling stories of people creating the change that can galvanize communities into action. When I heard that Washington State University, Shared Strategy for Puget Sound, and Farming and the Environment would join Sustainable Northwest in this enterprise, I was convinced that this project would be a success.

I want to take this opportunity to specially recognize and commend the work of Sustainable Northwest since this book is an extension of their decade long efforts to find, catalog, disseminate and network the most remarkable "change agents" in the Pacific Northwest. In a few short years, Sustainable Northwest has enabled communities, enterprises, and individuals to embrace new paths that pursue environmental conservation and restoration,

rebuild viable economic opportunities in rural areas, and meaningfully involve multiple stakeholder and communities of place and interest in defining their shared interests. Simply stated, this is the only way to the future of sound natural resource management with economically viable communities.

The National Fish and Wildlife Foundation is very pleased to have been a partner by providing the seed financial support for this project with the U.S. Fish & Wildlife Service through our Washington State Community Salmon Fund. We hope the stories contained in this volume are widely distributed and used by teachers, landowners and businesses, community leaders and elected officials, as well as government agencies. Most importantly, I want to thank the people in these stories for their remarkable achievements, strength, wisdom, commitment and inspiration for others to follow their examples.

Krystyna Wolniakowski, Director
National Fish and Wildlife Foundation-Northwest
Portland, Oregon

ACKNOWLEDGEMENTS

Renewing the Countryside: Washington is the latest volume in a series inspired by a book published in the Netherlands and subsequently in many communities across the United States. We are happy to include the voices of Washington in a growing chorus of others who are renewing the countryside.

Renewing the Countryside: Washington was initially funded with a catalytic grant from the National Fish and Wildlife Foundation and its Community Salmon Fund. We wish to thank Krystyna Wolniakowski, Pacific Northwest Director, for her abiding support for our work, as well as her colleagues Jennifer Taylor and Collette Lord. We also wish to thank the Northwest Area Foundation for making an early purchase order to distribute 1000 copies of this book. Special thanks go to Karl Stauber, its President, Community Liaison Karla Miller and to former Program Lead, Patrick Murphy. In addition, financial assistance was generously provided by Sam Walton through the Walton Family Foundation.

John Harrington has patiently and effectively managed this project over the past year. He conducted field visits and interviews, coordinated with publishing partners, and spent many evenings and weekends editing stories and selecting photos. The story selection and editorial committee included Richard Hines (Washington State University Center for Sustaining Agriculture & Natural Resources), Jagoda Perich-Anderson (Shared Strategy for Puget Sound), Jeanne Wallin (Farming and the Environment), and John Harrington (Sustainable Northwest). Theirs was a very tall order indeed, for it was very difficult to choose from the scores of inspiring and remarkable story submittals.

Ingrid Dankmeyer transcribed hours of interviews, authored the stories and chapter headings, and assisted with editing. We are most grateful to Ingrid for her adoption of this project and the singular dedication she gave to it. We also thank her husband, Pete Dubois, for his editing and support.

Many talented photographers contributed to this publication. Thanks to James Anderson who traveled many miles to capture an impressive array of images. His talents, flexibility and enthusiasm have been invaluable. Thanks also to photographers Sam Walton, David Perry, and Ashlyn Forshner for their photo contributions to this publication.

Betty Oppenheimer generously allowed us to quote excerpts from her book, "Growing Lavender and Community on the Sequim Prairie" for the Sequim Lavender Festival profile. We also want to extend a special thanks to Matthew Buck for his initiation of this project while working at Sustainable Northwest.

This project was completed in partnership with Renewing the Countryside who provided guidance in its development and implementation. Renewing the Countryside's work on this project was generously supported by the W. K. Kellogg Foundation.

A very special thanks to the public agencies, native tribes, nonprofit organizations, community groups, and individuals who contributed to the success of these stories, but are not recognized in this book. Their "behind the scenes" dedication and tenacity are often the driving force behind the sustainable restoration and revitalization of Washington's natural resources and rural communities.

Martin Goebel, President, Sustainable Northwest

PUBLISHING PARTNERS

Sustainable Northwest

Sustainable Northwest partners with communities and enterprises to achieve economic, ecological, and community vitality and resilience. We work to foster an economy and society in the Pacific Northwest where people, communities, and businesses refuse to sacrifice the good of the land for the good of the people, or the good of the people for the good of the land – finding a new path which honors both. We are committed to a human community working together - able to think beyond itself to embrace the entire biological community and from one generation to many.

We believe that people are an indivisible part of the ecosystems they inhabit; economic and environmental health are interdependent; communities have the energy and creativity to develop innovative, lasting solutions to complex environmental, economic, and social challenges; effective ecosystem stewardship is adaptive, place-based and founded on evolving scientific and practical knowledge; and connections between rural and urban communities and collaboration between diverse interests and individuals are integral to local and regional sustainability.
For more information about our programs visit:
www.sustainablenorthwest.org.

Washington State University Center for Sustaining Agriculture and Natural Resources

Washington State University's Center for Sustaining Agriculture and Natural Resources (CSANR) fosters approaches to agriculture and natural resource stewardship that are economically viable, environmentally sound, and socially responsive.

Our state is known for its natural beauty—highly varied landscapes and numerous microclimates. These same conditions allow the state to produce an impressively diverse range of farm goods. Citizens here support a vibrant food and farming system, and they look to public institutions like WSU to help address the challenges now facing agriculture.

CSANR helps find solutions by connecting the land-grant institution with rural and urban communities across the state. CSANR is guided by an interdisciplinary Leadership Team, which identifies priorities for action and develops unique projects. As a result, Washington farms are more viable and our citizens enjoy better access to fresh, healthy, local food. Learn more at: www.csanr.wsu.edu and send us your ideas at csanr@wsu.edu.

Shared Strategy for Puget Sound

The Shared Strategy for Puget Sound is a collaborative initiative to recover and maintain an abundance of naturally spawning salmon at harvestable levels. The Shared Strategy engages farmers, timberland·owners, fishers, tribes, developers and all levels of government to create a future in which both people and salmon co-exist and thrive.

The Shared Strategy believes that an investment in salmon is an investment in our state's economy. Salmon are a symbol of our region's health—when they decline, our quality of life and related economic drivers, such as tourism, recreation and fisheries are at risk. People in Washington state are no longer satisfied with polarized either/or choices and are increasingly helping search for, find and implement "both/and" solutions that help the environment and their communities' social and economic well-being. The Shared Strategy works with its partners and stakeholders to identify such innovative solutions and has collected a wealth of examples, many of which will find their way into the Puget Sound salmon recovery plan. Learn more at: www.sharedsalmonstrategy.org.

Farming and the Environment

Farming and the Environment was formed to protect the environment and enhance the economic viability of agriculture in Washington State by providing farmers and ranchers the support and resources they need to be good stewards of Washington's air, land and water.

To achieve our mission, Farming & the Environment has been launching programs to recognize and reward farmers who are exemplary stewards of the land; raise public awareness about the connection between how food is grown and its quality, and how everyday food choices help protect our farmlands. We're designing these programs with help from scientists and scholars at Washington State University who are researching solutions to environmental and economic challenges currently facing farmers and ranchers. To learn more visit: www.farmingandtheenvironment.org.

Renewing the Countryside

Renewing the Countryside (RTC) is a Minnesota-based non-profit organization that works with partners from across the country to collectively share stories of people who are redefining what it means to live, work, and learn in rural America. These stories provide hope, inspiration, and ideas for building strong, sustainable rural communities.

Through the generous support of individuals and foundations, Renewing the Countryside is dedicated to sharing the strength of America's rural landscape: the people enhancing their cultural and natural resources while spurring local economic development in their communities.

To find out about other Renewing the Countryside publications or to read more stories of people revitalizing their rural communities, visit the Renewing the Countryside website at www.renewingthecountryside.org.

Story Contacts

Community

1. Dungeness River Watershed Restoration
   Ann E. Seiter
   PO Box 4
   Curlew, WA 99118

2. Left Foot Organics
   Ann M. Vandeman, Executive Director
   P.O. Box 12772
   Olympia, WA 98502
   Tel: 360-754-1849
   info@leftfootorganics.org
   www.leftfootorganics.org

3. Made in the Methow Co-op
   108 N. Glover Street
   Twisp, WA 98856
   Tel: 509-997-7482

   Made in the Methow Co-op
   Ann Wagstaff, Kitchen Coordinator
   PO Box 116
   Carlton, WA 98814
   Tel: 509-997-5420
   Fax: 509-997-5420

4. Othello Sandhill Crane Festival
   PO Box 542
   Othello, WA 99344
   www.othellosandhillcranefestival.org

5. Sequim Lavender Festival
   105 1/2 E First St.
   Port Angeles, WA 98362
   Tel: 360-681-3035 or 877-681-3035
   Fax: 360-452-4695
   www.lavenderfestival.com

   Sequim Lavender Festival
   Betty Oppenheimer
   10 Salal Way
   Sequim, WA 98382
   Tel: 360-683-3441
   ravensbop@olympus.net
   Betty Oppenheimer is the author of "Growing
   Lavender and Community on the Sequim Prairie:
   A How-to and History." The book includes a history
   of Sequim's lavender industry, and information on
   growing, harvesting, and making crafts from
   lavender. Available direct from the author for
   $19.95 plus shipping and handling.

6. Tenmile Creek Watershed Restoration Project
   Dorie Belisle, Project Manger
   231 Ten Mile Rd.
   Lynden, WA 98264
   Tel: 360-398-9187
   doriebelisle@yahoo.com
   www.bellewoodapples.com
   www.whatcomcd.org

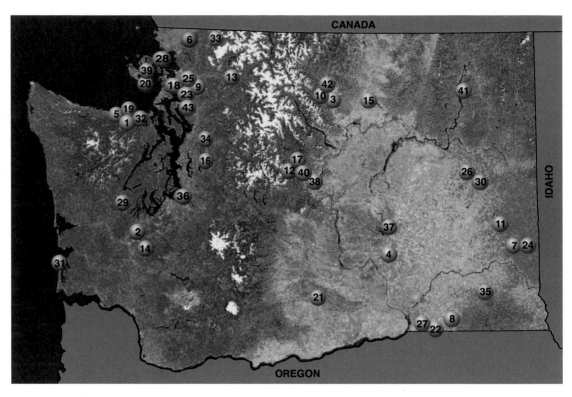

Conservation

7. JEA Farms, Ltd.
   John Aeschliman
   201 Aeschliman Rd.
   Colfax, WA 99111
   jeaesc@colfax.com

8. Leonetti Cellar LLC
   Chris Figgins
   1875 Foothills Lane
   Walla Walla, WA 99362
   Tel: 509-525-1428
   www.leonetticellar.com

9. Limberlost Tree Farm
   Herb & Grace Payne
   9355 Reef Point Lane
   LaConner, WA 98257
   Tel: 360-466-4452
   grace.payne@verizon.net

10. Moccasin Lake Ranch
    Winthrop, WA

11. Read, Deanna & Jeremy Smith Family Farms
    J. Read Smith
    11751 Lancaster Rd.
    St. John, WA 99171
    Tel: 509-648-3922
    Fax: 509-648-3922
    reads@stjohncable.com

12. Two Rivers Farm
    Nancy Denson & Nick Stemm
    12450 Wilson Street
    Leavenworth, WA 98826
    Tel: 509-548-4422
    ndenson@tworiversfarm.com

Farming and Ranching

13. Blue Heron Farm & Nursery
    Anne Schwartz
    12179 State Route 530
    Rockport, WA 98283
    Tel: 360-853-8449
    als@fidalgo.net
    www.blueheronfarmandnursery.com

14. Colvin Family Ranch
    Fred A. Colvin
    16816 Old Hwy. 99 SE
    Tenino, WA 98589
    Tel: 360-264-2890
    fkcolvin@cs.com

15. Double J Ranch
    Peter Goldmark
    400 Timentwa Rd.
    Okanogan, WA 98840

16. Full Circle Farm
    Andrew Stout
    PO Box 608
    Carnation, WA 98014
    Tel: 425-333-4677
    Fax: 425-333-4678
    info@fullcirclefarm.com
    www.fullcirclefarm.com

17. Gibbs' Organic Produce
    Grant Gibbs
    11632 Freund Canyon Road
    Leavenworth, WA 98826

18. Hedlin Farms
    Dave Hedlin
    12275 Valley Road
    Mt. Vernon, WA 98273
    Tel: 360-466-3977
    hedlin@hedlinfarms.com

19. Dungeness Organic Produce
    Nash Huber, Owner
    Kia I. Kozun, Operations Manager
    1865 East Anderson Road
    Sequim, WA 98382
    Tel: 360-681-7458
    Fax: 360-683-6807
    nashsorganicgirl@yahoo.com
    www.nashsproduce.com

20. S & S Homestead
    Henning Sehmsdorf
    2143 Lopez Sound Road
    Lopez Island, WA 98261
    360-468-3335
    sshomestead@rockisland.com
    http://csanr.wsu.edu/educationopps/internships.htm

BUSINESS
21. Inaba Produce Farms, Inc.
    Lon Inaba
    8351 McDonald Rd.
    Wapato, WA 98951
    Tel: 509-848-2982

22. Locati Farms, Inc.
    Michael Locati
    PO Box 327
    Walla Walla, WA 99362
    Tel: 509-525-0286
    Fax: 509-525-9595
    locatif@pocketinet.com
    www.locatifarms.com

23. Mike and Jean's Berry Farm
    Mike & Jeanne Youngquist
    16402 Jungquist Rd.
    Mount Vernon, WA 98273
    Tel: 360-424-7220
    Fax: 360-424-7225
    mjberry@fidalgo.net
    www.mikeandjeans.com

24. Paradise Fibers
    Kate Painter
    701 Parvin Rd
    Colfax, WA 99111
    paradise_fibers@yahoo.com
    www.paradisefibers.com

25. Sakuma Bros. Holding Company
    Steven M. Sakuma
    PO Box 427
    Burlington, WA 98233
    Tel: 360-757-6611
    steves@sakumabros.com

26. Shepherd's Grain
    Columbia Plateau Producers, LLC
    Karl Kupers
    12996 Kupers Road North
    Harrington, WA 99134
    Tel: 509-721-0374
    karl@shepherdsgrain.com
    www.shepherdsgrain.com

    Shepherd's Grain
    Columbia Plateau Producers, LLC
    Fred J. Fleming
    29768 State Route 231 N
    Reardan, WA 99029
    Tel: 509-979-1132
    Fax: 509-796-2576
    fredfj2@aol.com
    www.shepherdsgrain.com

27. Thundering Hooves Family Farm
    Joel Huesby
    1511 Fredrickson Road
    Touchet, WA 99360
    Tel: 866-350-9400
    Fax: 509-522-9444
    www.thunderinghooves.net

28. Willows Inn, Inc.
    Riley D. Starks
    2579 West Shore Drive
    Lummi Island, WA 98262
    Tel: 360-758-2620
    Fax: 360-758-7399
    riley@willows-inn.com
    www.willows-inn.com
    www.nettlesfarm.com
    www.lummiislandwild.com

RESTORATION
29. Goldsborough Dam Removal
    Simpson Timber Company
    www.simpson.com

    Goldsborough Dam Removal
    Squaxin Island Tribe
    Jeffrey A. Dickison
    2952 SE Old Olympic Hwy
    Shelton, WA 98584
    Tel: 360-432-3815
    Fax: 360-426-3971
    jdickison@squaxin.nsn.us
    www.squaxinisland.org

30. Healing Hooves, LLC
    Craig Madsen
    PO Box 148
    Edwall, WA 99008
    Tel: 509-990-7132
    Fax: 509-236-2451
    shepherd@healinghooves.com
    www.healinghooves.com

31. Hines Marsh Restoration
    The Trumpeter Swan Society
    Martha Jordan, Boardmember
    PMB 272
    914- 164th Street SE
    Millcreek, WA 98012
    www.trumpeterswansociety.org

32. JimmyComeLately Creek and Estuary Restoration
    Jamestown S'Klallam Tribe
    1033 Old Blyn Highway
    Sequim, WA 98382
    Tel: (360) 683-1109
    Fax: (360) 681-3405
    info@jamestowntribe.org
    www.jamestowntribe.org

33. Nuqua'lum
    Gary & Ellen Gehling
    PO Box 654
    Maple Falls, WA 98266
    Tel: 360-599-3425
    nuqualum@gte.net
    www.nuqualum.com

34. Skykomish River and Haskell Slough Restoration
    Northwest Chinook Recovery
    John A. Sayre, Executive Director
    15657 Yokeko Drive
    Anacortes, WA 98221
    Tel: 360-588-1917
    nwchinook@aol.com

35. Tucannon River Restoration
    Donald Howard
    1420 Tucannon Rd.
    Pomeroy, WA 99347

INNOVATIONS
36. Forest Concepts, LLC
    James H. Dooley
    Tel: 253-838-4759
    jdooley@seanet.com
    www.forestconcepts.com

37. Gies Farms, Inc.
    Dale Gies
    11653 Rd. 5 SE
    Moses Lake, WA 98837
    Tel: 509-765-9672
    djgies@atnet.net

38. IronStraw Group
Michael and Spring Thomas
P.O. Box 401
Oakland, OR 97462
[formerly of Cashmere, WA]
Tel: 541-817-5156
thomas@ironstraw.org
www.ironstraw.org

39. Lopez Mobile Processing Unit
Island Grown Farmers Cooperative
Bruce Dunlop
Lopez Island Farm
193 Cross Rd.
Lopez Island, WA 98261
Tel: 360-468-4620
bruce@lopezislandfarm.com

40. Peshastin Creek Growers Association
Dennis Nicholson
P.O. Box 55
Peshastin, WA 98847
Tel: 509-548-4207
Fax: 509-548-2092
nichorch@rightathome.com
www.ourorchard.com

Peshastin Creek Growers Association
John Dunley
Washington State University
Tree Fruit Research and Extension Center
dunleyj@wsu.edu
http://entomology.tfrec.wsu.edu/pearent

41. Quillisascut Farm
Richard M. & LoraLea Misterly
2409 Pleasant Valley Rd.
Rice, WA 99167

Tel: 509-738-2011
loralea@quillisascutcheese.com
www.quillisascutcheese.com

42. Rendezvous Reclamation
Sam Lucy
PO Box 1082
Winthrop, WA 98862
Tel: 509-996-3526
sam@methownet.com

43. Stillaguamish Old Channel Restoration and Tidegates
Stillaguamish Flood Control District
Chuck Hazleton
23224 Marine Dr.
Stanwood, WA 98292
Tel: 360-652-9233
Fax: 360-652-2494
chaz@snohomish.net

PHOTOGRAPHY CREDITS